Probability and Mathematical Statistics (

WILLIAMS • Diffusions, Markov
Foundations

ZACKS • Theory of Statistical Inf

Applied Probability and Statistics

ANDERSON, AUQUIER, HAUCK, OAKES, VANDAELE, and WEISBERG • Statistical Methods for Comparative Studies

ARTHANARI and DODGE • Mathematical Programming in Statistics

BAILEY • The Elements of Stochastic Processes with Applications to the Natural Sciences

BAILEY • Mathematics, Statistics and Systems for Health

BARNETT • Interpreting Multivariate Data

BARNETT and LEWIS • Outliers in Statistical Data

BARTHOLOMEW • Stochastic Models for Social Processes, *Third Edition*

BARTHOLOMEW and FORBES • Statistical Techniques for Manpower Planning

BECK and ARNOLD • Parameter Estimation in Engineering and Science

BELSLEY, KUH, and WELSCH • Regression Diagnostics: Identifying Influential Data and Sources of Collinearity

BENNETT and FRANKLIN • Statistical Analysis in Chemistry and the Chemical Industry

BHAT • Elements of Applied Stochastic Processes

BLOOMFIELD • Fourier Analysis of Time Series: An Introduction

BOX • R. A. Fisher, The Life of a Scientist

BOX and DRAPER • Evolutionary Operation: A Statistical Method for Process Improvement

BOX, HUNTER, and HUNTER • Statistics for Experimenters: An Introduction to Design, Data Analysis, and Model Building

BROWN and HOLLANDER • Statistics: A Biomedical Introduction

BROWNLEE • Statistical Theory and Methodology in Science and Engineering, *Second Edition*

BURY • Statistical Models in Applied Science

CHAMBERS • Computational Methods for Data Analysis

CHATTERJEE and PRICE • Regression Analysis by Example

CHERNOFF and MOSES • Elementary Decision Theory

CHOW • Analysis and Control of Dynamic Economic Systems

CHOW • Econometric Analysis by Control Methods

CLELLAND, BROWN, and deCANI • Basic Statistics with Business Applications, *Second Edition*

COCHRAN • Sampling Techniques, *Third Edition*

COCHRAN and COX • Experimental Designs, *Second Edition*

CONOVER • Practical Nonparametric Statistics, *Second Edition*

CORNELL • Experiments with Mixtures: Designs, Models and The Analysis of Mixture Data

COX • Planning of Experiments

DANIEL • Biostatistics: A Foundation for Analysis in the Health Sciences, *Second Edition*

DANIEL • Applications of Statistics to Industrial Experimentation

DANIEL and WOOD • Fitting Equations to Data: Computer Analysis of Multifactor Data, *Second Edition*

DAVID • Order Statistics, *Second Edition*

DEMING • Sample Design in Business Research

DODGE and ROMIG • Sampling Inspection Tables, *Second Edition*

DRAPER and SMITH • Applied Regression Analysis, *Second Edition*

DUNN • Basic Statistics: A Primer for the Biomedical Sciences, *Second Edition*

DUNN and CLARK • Applied Statistics: Analysis of Variance and Regression

ELANDT-JOHNSON • Probability Models and Statistical Methods in Genetics

ELANDT-JOHNSON and JOHNSON • Survival Models and Data Analysis

continued on back

Discrimination and Classification

Discrimination and Classification

D. J. Hand
Biometrics Unit,
London University, Institute of Psychiatry

JOHN WILEY & SONS
Chichester · New York · Brisbane · Toronto

Library of Congress Cataloging in Publication Data

Hand, D. J.
Discrimination and classification.—(Wiley series in probability and mathematical statistics).
Bibliography: p.
Includes index.
1. Discriminant analysis. I. Title.
QA278.65.H36 519.5′35 81-13045
ISBN 0 471 28048 8 AACR2

British Library Cataloguing in Publication Data:

Hand, D. J.
Discrimination and classification.—(Wiley series in probability and mathematical statistics).
1. Discriminant analysis
I. Title
519.5′3 QA278

ISBN 0 471 28048 8

Typeset by Preface Ltd, Salisbury, Wilts.,
and printed in the United States of America.

Contents

Preface

In the discussion following a paper on error rate estimation (Hills, 1966) Wagle said:

> The subject of discriminant analysis has developed through three stages. The first was the Fisherian stage using the intuitive approach and developing the theory of linear discriminant functions. This was followed by the probabilistic stage considered by Welch, Rao, and others. The third stage was the Waldian stage based on the principles of statistical decision theory. All these stages basically assumed underlying multivariate normal populations when dealing with numerical problems, and replaced the unknown parameters by the sample estimates. The subject of non-parametric discrimination has received very little attention.

Now we have passed through a fourth and perhaps a fifth stage. Non-parametric methods of discriminant analysis have received considerable attention and recent years have also witnessed a tremendous growth in methods of tackling classification and cluster analysis problems. But perhaps more important has been the growth in interest in discrimination and classification problems from areas outside traditional statistics. The particular value of such developments lies in the new viewpoints and different emphases which have resulted. These in turn can lead to improved understanding and novel applications.

This book is an attempt to collect together the different approaches to discrimination and classification problems in a form suitable for statisticians. Although it is not possible to be exhaustive, I hope the bibliography will serve as a guideline and introduction to the main areas of the literature.

The level of mathematics is such that undergraduates in statistics should have little difficulty in following it. The exercises after each chapter are a mixture of computer projects (where I have tried to concentrate on the most important and widely used algorithms), numerical examples, and theoretical questions. The latter further develop ideas introduced in the text.

I am indebted to my colleague, Mr Brian Everitt, for his constructive comments on an early draft, and for giving me the benefit of his own experience as an author. I am also especially grateful to Mrs Bertha Lakey, who transformed my handwritten drafts into legible typescript with exceptional efficiency.

London University
Institute of Psychiatry
1981

D. J. HAND

CHAPTER 1

Introduction

1.1 SCOPE AND BACKGROUND

In recent years, chiefly as a consequence of the development of the electronic computer, considerable advances have been made in the practical application of statistical classification techniques. The classical work, beginning with Fisher (1936), has been supplemented by efforts to model and even improve upon human classification abilities. This work, going under the name of pattern recognition, has seen publication mainly in computer science, electrical engineering, and artificial intelligence journals. The aim of this book is to summarize these advances and relate them to the statistical work to produce a unified overview of discrimination and classification methods.

The central problem with which we are concerned is that of assigning an object into one of several possible categories. Couched in such terms it is immediately apparent that the range of application of the techniques to be outlined here will be very broad, and this is indeed the case. As illustrative examples of this breadth, and to give a more concrete idea of the type of problem, consider the following.

(a) One of the earliest applications of techniques of this kind was in the classification of an archaeological specimen—a skull—to one of two races: English or Eskimo.

(b) In contrast, one of the most recent applications has been to the classification of crops from high altitude photographs. It can be cheaper to estimate total acreage this way than by ground measurement, and it is also possible to identify incipient crop disease before it can be recognized from the ground. Similar methods have been applied to the detection of mineral deposits.

(c) Medical diagnosis is a particularly fruitful area of application. The methods have been applied to the assessment of the prognostic value of tests of lung function in miners with pneumoconiosis, to selecting appropriate operations for breast cancer patients, to predicting ischaemic heart disease, to predicting relapse in pulmonary tuberculosis sufferers, etc. The list is virtually endless.

(d) Amongst the recent applications which will have a very direct effect on our everyday life are those of speech recognition and optical character recognition.

(e) Speech recognition is an example of a case where the objects to be classified are waveforms. Other such examples are cardiac wave analysis, target recognition from vehicle noise or radar returns, and automatic electro-encephalograph analysis.

(f) Personnel classification is an area of increasing though perhaps controversial importance. The methods have been applied in vocational guidance, academic attainment, and even in recognizing high-risk individuals in psychiatric research.

(g) Other areas of application include anthropology, ecology, taxonomy, author verification, psychology, fingerprint recognition, etc. Lachenbruch (1975) lists 579 references.

Although it may not be apparent from the above examples, in fact there are two types of problem, one of which in a sense logically precedes the other.

(a) Class definition: the problem of imposing a classification scheme on the objects.

(b) Allocation: the problem of devising a classification rule from samples of already classified objects.

Techniques for tackling (a), that is grouping objects into classes according to their 'similarity', have gone under many names, but the most popular seems to be *cluster analysis*. The methods are applicable when no division of the objects into categories is available. For problem (b) the existence of classified samples implies that some kind of classification scheme exists, but it may be subjective, not implementable on a digital computer, retrospective (for example, predicting the outcome of cancer operations), or unusable for other reasons. Somehow the essence of the classification scheme must be extracted and transformed into a practical classification rule. Thus, although it is true that class definition is necessary before classification rules can be devised and, hence, new objects classified, for us the problems are distinct rather than consecutive.

For both problems our aim in mathematical terms is to find a function mapping objects to an index set consisting of class identifiers. The term *discrimination* refers to the process of deriving classification rules from samples of classified objects, while the term *classification* refers to applying the rules to new objects of unknown class (we should point out that some authors use these terms with different meanings, but no confusion should arise since the context usually makes the intended meaning clear).

It will be manifest to the reader that for some discrimination and classification problems the human perceptions are pre-eminent. In determining sex merely from facial characteristics or in determining nationality from accent, for example, human abilities would be hard to beat. However, in other areas this is not so. This is especially the case when the objects cannot be perceived directly and they are represented by numerical data in tabular form. Few humans can glance down an array of numbers (with, say, one row of numbers for each object) and group the objects into classes. It is possible to convert such raw data into more easily grasped forms (see, for instance, Chernoff, 1973, for a description of representing multivariate data as cartoon faces) but this still presents the problem that any grouping or classification based on these diagrams is subjective.

This suggests the following reasons for developing formal statistical classification methods:

(i) They are objective and can be repeated by other researchers.

(ii) We can assess the performance of the assignment rule.

(iii) We can formally measure the relative sizes of the classes.

(iv) We can determine how representative is a particular example of its class.

(v) We can investigate what aspects of the objects are important in producing a classification.

(vi) We can describe and test the differences between classes.

But even for those areas where humans can perform adequately, there are advantages in statistical methods:

(vii) The speed of classification is greater by computer (important, for example, in identifying a particular particle track amongst hundreds of thousands of bubble chamber photographs, or in classifying each picture element in a 1000 by 1000 element satellite photograph).

(viii) The burden of boring and repetitive classification tasks can be eased by automation.

1.2 THE MEASUREMENT SPACE

The methods to be outlined in this book are based on the usual multivariate statistical representations. That is to say, each object is 'transformed' by having a number of measurements taken on it and each measurement provides one variable in a multidimensional space. The statistical techniques deal with the vectors or points representing objects in this multidimensional space. The generality of the methods derives from the abstract nature of the vector representation, but clearly exactly what is measured will depend on the particular problem. For example:

(a) in trying to classify a surgical wound into one of two classes (will/will not turn septic) we could use patient age, duration of operation, length of incision, etc.;

(b) in machine classification of spoken words we will take measurements characterizing the waveforms of the words; and

(c) in personnel classification we could take as measurements the responses to various psychological tests and questionnaires.

Usually there are an unlimited number of different measurements which could be taken, but there will exist complex relationships between them so that selecting a good set of measurements is not a trivial exercise. Paradoxical though it may seem, there can be penalties in increased error (as well as cost) in using too many measurements as well as in using too few. For this reason techniques have been developed for choosing (by selection or transformation) small sets of measurements from the initial possible large set. In formal pattern recognition terminology the elements of the initial set are known as *measurements*, together constituting the *measurement space*, while the elements of the extracted small set are known as *features*, forming the *feature space*. We shall use the word *variable* to describe both sets—no confusion will arise since it will be obvious from the context to which set we are referring. In fact, throughout most of the book we shall be using the extracted feature set, the exception being Chapter 6 where we actually discuss methods for finding good feature sets from measurements.

Since the objects are transformed into vectors and represented by points in a multidimensional space it makes sense to talk formally about distances between objects. Our hope, of course, is that similar objects should lead to points which lie near each other. Our aim is thus to define regions of the space so that new points (and their corresponding objects) can be classified according to the region into which they fall.

1.3 DECISION THEORY

In order to introduce the theoretical basis underlying discrimination and classification we shall need some definitions. A *decision rule* partitions a space into regions Ω_i, $i = 1, \ldots, N$, where N is the number of classes (each of which may be multiply connected). An object is classified as coming from class ω_k if its corresponding vector representation, $\mathbf{x}$, lies in region Ω_k. (Readers familiar with game theory will recognize this as a *deterministic* decision rule. Since our 'opponent', Nature, cannot alter her strategy, there is no advantage to be gained from *randomized* decision rules.) The boundaries between regions are called *decision surfaces*.

We shall begin by assuming that we know the probability $P(\omega_i)$ that an object comes from class ω_i ($i = 1, \ldots, N$). Since this is an overall probability, independent of $\mathbf{x}$, and known before we have taken any observations it is called a *prior probability*. If we had no further information then the best decision rule would be to classify an object as coming from class ω_k if

$$P(\omega_k) > P(\omega_i), \qquad i = 1, \ldots, N; \quad i \neq k$$

(Equal probabilities can be settled arbitrarily.) This rule minimizes the probability of making an error. Note that it is a rather trivial partitioning of the space, 'dividing' it into only one region and classifying all objects to class ω_k.

Usually, however, we do have further information, namely a vector, $\mathbf{x}$, of observations made on the object to be classified. (These observations may be direct measurements or they may be the result of a feature extraction process as mentioned in Section 1.2.) In this case we can compare the probabilities of belonging to each class *at* $\mathbf{x}$ and classify according to whichever is the larger

$$P(\omega_k|\mathbf{x}) > P(\omega_j|\mathbf{x}) \quad \text{for all } j \neq k \Rightarrow \mathbf{x} \in \Omega_k$$

(the symbol $\in$ means 'is an element of' and here signifies that $\mathbf{x}$ lies in Ω_k, i.e. that the object will be classified as belonging to class ω_k). This fundamental rule is known as *Bayes minimum error rule*. In contrast to the prior probabilities $P(\omega_i)$, the probabilities $P(\omega_i|\mathbf{x})$ depend on $\mathbf{x}$, that is they can only be calculated after the value of $\mathbf{x}$ has been determined. For this reason they are known as *posterior* (or *a posteriori*) probabilities. Unfortunately we rarely know these *a posteriori* probabilities, they will need to be estimated, and we can do this by making use of the samples of known classification. This is one of the approaches followed in later chapters. Often, however, a more convenient formulation of this rule is obtained by applying Bayes's theorem

$$P(\omega_i|\mathbf{x}) = \frac{p(\mathbf{x}|\omega_i)P(\omega_i)}{p(\mathbf{x})}$$

which yields

$$p(\mathbf{x}|\omega_k)P(\omega_k) > p(\mathbf{x}|\omega_j)P(\omega_j), \quad \text{for all } j \neq k \Rightarrow \mathbf{x} \in \Omega_k \tag{1}$$

(the $p(\mathbf{x})$ is common to both sides and can be omitted). If the class-conditional probability density functions (pdfs) $p(\mathbf{x}|\omega_i)$ are known then the problem is solved—we simply substitute the $\mathbf{x}$-vector, for the object to be classified, into (1) and find the largest value of $p(\mathbf{x}|\omega_i)P(\omega_i)$. But, as before, the $p(\mathbf{x}|\omega_i)$ are usually unknown and are estimated from the set of classified samples. In practice the distinction between methods based directly on $P(\omega_i|\mathbf{x})$ and those based on $p(\mathbf{x}|\omega_i)$ is more nominal than real, as will become apparent.

For two classes the Bayes minimum error decision rule can be expressed in the convenient likelihood ratio form

$$\frac{p(\mathbf{x}|\omega_1)}{p(\mathbf{x}|\omega_2)} \gtrless \frac{P(\omega_2)}{P(\omega_1)} \Rightarrow \mathbf{x} \in \begin{cases} \Omega_1 \\ \Omega_2 \end{cases}$$

Although this rule minimizes the overall error, in fact we might be interested in some other criterion. So far we have assumed that the penalty due to misclassifying a class ω_1 point as class ω_2 is the same as vice versa. But this is not always the case. Misclassifying a cancer sufferer as healthy is a much more serious error than the other way round. This concept has been formalized in terms of a *cost function*, C_{ij}, which is the cost of misclassifying an object from class ω_i as belonging to class ω_j. If $\mathbf{x} \in \omega_i$ the expected cost is

$$r_i = \sum_{j=1}^{N} C_{ij} \int_{\Omega_j} p(\mathbf{x}|\omega_i)\, d\mathbf{x}$$

The overall expected cost, or *risk*, is thus

$$r = \sum_i r_i P(\omega_i) = \sum_i \sum_j \int_{\Omega_j} C_{ij} P(\omega_i) p(\mathbf{x} | \omega_i)\, d\mathbf{x}$$

$$= \sum_j \int_{\Omega_j} \left\{ \sum_i C_{ij} P(\omega_i) p(\mathbf{x} | \omega_i) \right\} d\mathbf{x} \tag{2}$$

This will be minimized if we define Ω_k such that $\mathbf{x} \in \Omega_k$ whenever

$$\left\{ \sum_i C_{ik} P(\omega_i) p(\mathbf{x} | \omega_i) \right\} < \left\{ \sum_i C_{ij} P(\omega_i) p(\mathbf{x} | \omega_i) \right\} \quad \text{for all } j \neq k$$

This is the *Bayes minimum risk decision rule*. Note that if $C_{ii} = 0$ for all i and $C_{ij} = 1$ whenever $i \neq j$ this is the same as the Bayes minimum error rule.

For two classes this rule becomes

$$\frac{p(\mathbf{x} | \omega_1)}{p(\mathbf{x} | \omega_2)} \gtrless \frac{C_{21} - C_{22}}{C_{12} - C_{11}} \frac{P(\omega_2)}{P(\omega_1)} \Rightarrow \mathbf{x} \in \begin{cases} \Omega_1 \\ \Omega_2 \end{cases}$$

(assuming $C_{11} < C_{12}$).

When there are only two classes there are only two possible types of misclassification so we could base a decision rule on minimizing one, subject to a specified low value of the other. Let us begin by fixing the probability of misclassifying an element of ω_2 as being in ω_1 to be α, i.e.

$$\alpha = \int_{\Omega_1} P(\omega_2) p(\mathbf{x} | \omega_2)\, d\mathbf{x} \tag{3}$$

We now seek to minimize the probability of the other type of misclassification, defined by

$$\beta = \int_{\Omega_2} P(\omega_1) p(\mathbf{x} | \omega_1)\, d\mathbf{x}$$

subject to (3). We can do this by using the Lagrange multiplier λ

$$r = \int_{\Omega_2} P(\omega_1) p(\mathbf{x} | \omega_1)\, d\mathbf{x} + \lambda \left\{ \int_{\Omega_1} P(\omega_2) p(\mathbf{x} | \omega_2)\, d\mathbf{x} - \alpha \right\}$$

$$= (1 - \lambda\alpha) + \int_{\Omega_1} \lambda P(\omega_2) p(\mathbf{x} | \omega_2) - P(\omega_1) p(\mathbf{x} | \omega_1)\, d\mathbf{x}$$

(recalling that $\Omega_1 \cup \Omega_2$ is the whole space). This will be minimized if we choose Ω_1 such that

$$\mathbf{x} \in \Omega_1 \quad \text{if } \{\lambda P(\omega_2) p(\mathbf{x} | \omega_2) - P(\omega_1) p(\mathbf{x} | \omega_1)\} < 0$$

which leads again to the likelihood ratio

$$\frac{p(\mathbf{x} | \omega_1)}{p(\mathbf{x} | \omega_2)} \gtrless \frac{\lambda P(\omega_2)}{P(\omega_1)} \Rightarrow \mathbf{x} \in \begin{cases} \Omega_1 \\ \Omega_2 \end{cases}$$

Readers familiar with statistical hypothesis testing will recognize that a special case of this result is the Neyman–Pearson lemma. For specified forms for $p(\mathbf{x}|\omega_i)$ and $P(\omega_i)$ we can use (3) and the above definition of Ω_1 to estimate λ using numerical methods.

It sometimes happens that the prior probabilities—the relative probabilities of a new object coming from each class—are unknown. Since each of the three decision rules above uses the ratio of priors in the threshold this poses a problem, especially if the sample sizes for each class are not proportional to the class probabilities. The *minimax rule* is designed so that we minimize the maximum possible risk.

From (2) the risk r is

$$r = \sum_i P(\omega_i) \sum_j \int_{\Omega_j} C_{ij} p(\mathbf{x}|\omega_i)\, d\mathbf{x}$$

Suppose for simplicity that there are only two classes. Then r is a linear function of $P(\omega_1)$—and hence the maximum of r occurs when $P(\omega_1) = 0$ or $P(\omega_1) = 1$. The maximum is thus either

$$\sum_j \int_{\Omega_j} C_{2j} p(\mathbf{x}|\omega_2)\, d\mathbf{x} \quad \text{or} \quad \sum_j \int_{\Omega_j} C_{1j} p(\mathbf{x}|\omega_1)\, d\mathbf{x}$$

If we take the particular cost function with $C_{11} = C_{22} = 0$ then the maximum of r becomes either

$$\int_{\Omega_1} C_{21} p(\mathbf{x}|\omega_2)\, d\mathbf{x} \quad \text{or} \quad \int_{\Omega_2} C_{12} p(\mathbf{x}|\omega_1)\, d\mathbf{x}$$

In view of the reciprocal relationship between these two risk values (since $\Omega_1 \cup \Omega_2$ is the complete space) we have

$$\max\left(\int_{\Omega_1} C_{21} p(\mathbf{x}|\omega_2)\, d\mathbf{x}, \quad \int_{\Omega_2} C_{12} p(\mathbf{x}|\omega_1)\, d\mathbf{x}\right)$$

takes its minimum value when

$$\int_{\Omega_1} C_{21} p(\mathbf{x}|\omega_2)\, d\mathbf{x} = \int_{\Omega_2} C_{12} p(\mathbf{x}|\omega_1)\, d\mathbf{x}$$

If $C_{21} = C_{12}$ the minimax rule is thus to choose the regions Ω_1 and Ω_2 so that the probabilities of the two types of error are the same. Although this book deals primarily with those decision rules which can be expressed in the form of a ratio of probabilities, extensions to rules such as the minimax will be straightforward.

We have expressed the rules above as functions of $\mathbf{x}$ via the class-conditional pdfs $p(\mathbf{x}|\omega_i)$. However, in those rules the absolute values of the probabilities do not matter—we are merely concerned with their relative magnitudes. (Is $p(\omega_k|\mathbf{x}) > p(\omega_j|\mathbf{x})$? Or is $p(\omega_1|\mathbf{x})/p(\omega_2|\mathbf{x}) >$ some constant?) Sometimes advantage can be taken of this fact. We can now write the rules more generally as

$$q_k(\mathbf{x}) > q_j(\mathbf{x}) \quad \text{for all } j \neq k \Rightarrow \mathbf{x} \in \Omega_k$$

or, for the two-class case

$$h(\mathbf{x}) \gtrless \text{constant} \Rightarrow \mathbf{x} \in \begin{cases} \Omega_1 \\ \Omega_2 \end{cases}$$

These general functions, q and h, are called *discriminant functions*. Note that any monotonic function of q or h could replace them without affecting the decision rule. This point is developed further in Chapter 4.

As stated before, the class-conditional pdfs (or the discriminant functions) will usually be unknown and will need to be estimated from the classified samples. Chapters 2 and 3 deal with estimating these pdfs under various levels of assumption, and Chapter 4 considers a particular type of discriminant function—the linear discriminant function—which has computational advantages and which, because of its analytic form, has received considerable theoretical development.

1.4 DESIGN SETS AND TEST SETS

In the preceding section we discussed how decisions could be based on probabilities estimated from a set of objects with known classifications. This set is known as the *design set* for the simple reason that the decision rule is 'designed' from this set, i.e. the defining parameters of the decision rule are estimated from this set (for example, in Chapter 3) or new points are classified by comparison with the members of this set (for example, in Chapter 2). This set is also sometimes called the *training set*, a name arising from pattern recognition work involving adaptive or sequential estimation (see Chapter 4).

The rules discussed in the preceding section are ideal rules, and when we replace the pdfs (or discriminant functions) in them by estimates we must expect the rules not to perform as well as they would if we knew the pdfs. This is not only because our assumptions might be wrong (for example, for practical convenience we may have assumed linear discriminant functions or normal class-conditional pdfs) but also because our estimates will be based on the design set, which is necessarily finite. Since each rule is designed by optimizing some performance criterion, we are simply saying that practical rules will fall short of the theoretically optimum value of the criterion.

In practice each of the rules discussed in the preceding section represents many different rules since they may be implemented using different approaches (for example, non-parametric versus parametric pdf estimates) and different estimation methods. For rules which optimize the same criterion we can choose between them by comparing the values of the criterion. For example, a choice between non-parametric and parametric pdf estimates could be made by seeing which approach led to the lowest proportion of misclassifications (termed *misclassification rate* or *error rate*).

Now a word of caution, which we shall outline in terms of the misclassification rate, but which applies in an analogous way to any other criterion. To compare rules on the basis of the misclassification rate a simple way would be to compare the number of design set points which the rules misclassify. This was indeed the approach adopted by early workers in pattern recognition. However, it was quickly discovered that the estimated error rates resulting from such a method were over-optimistic—when the rules were applied to new sets of data the error rates were greater than had been expected. In retrospect this result is obvious. After all, the decision rules are optimized on the design set—their parameters are estimated to minimize the design set misclassification rate (or some criterion related to it). Naturally, a new set, which is a random sample (from the same distribution), will be different from the design set and the parameters will not be quite optimal for this particular set.

Although these points are generally true it will be seen that the larger the design set the more representative it will be of the underlying population: so the nearer to the theoretical optimum will be the decision rule based on that set (subject to any constraints imposed by assumptions about the forms of pdfs or discriminant functions, of course).

To return to the original point, which was that of estimating the error rates of decision rules for comparison purposes, if the design set cannot be used, what alternatives are there? One approach is to use an independent *test set* of data—a set of objects, again of known classification. The decision rule derived from the design set is assessed by classifying the test-set elements.

A striking example of the phenomenon of optimistic design set misclassification rates was provided during the course of developing a nine-item mental health screening questionnaire. On the design set a misclassification rate of 0.6 per cent was obtained. But on the independent test set a rate of 21.2 per cent was obtained.

The optimistic error rate estimate based on reclassifying the design set is known as the *apparent error rate*. The rate resulting from classifying an independent test set is an estimate of the *true error rate* for that decision rule. It should be noted, however, that splitting the data set into design and test sets in this way does not make the best use of it. A more reliable decision rule would result if it were based on the entire data set. With this in mind other evaluation methods have been devised, and these are discussed in Section 8.1. If some criterion other than simple error rate is adopted as a basis for comparison the arguments are direct analogous to those above.

1.5 DISTANCES

Since our aim is to partition the space containing the vector representations of the objects into regions, each region ideally containing objects from only a single class, the concept of distance or dissimilarity between objects is an important one for us. The distance measure with which we are most familiar is the *Euclidean metric*, defining the distance between the vector representations **x** and **y** (where the components are interval scales) by

$$d(\mathbf{x}, \mathbf{y}) = \left\{ \sum_i (x_i - y_i)^2 \right\}^{1/2}$$

This measure has certain advantages, in particular analytic ones, since it possesses a derivative at every point. However, we are not forced to use this measure and it may be that others have different advantages in some situations. Moreover, in some cases the requirement of the Euclidean metric that the scales have at least interval properties might make it inappropriate. Nevertheless most of the work on discriminant analysis—i.e. work on finding classification rules from classified samples—has been in terms of the Euclidean metric. In cluster analysis, on the other hand, considerable attention has been given to the choice of distance measure. For these reasons we have postponed the detailed discussion of alternative measures to Chapter 7.

Unfortunately, even if one decides to use the Euclidean metric there remain questions. As an example consider the (somewhat artificial) classification problem involving classifying towns on the basis of two measurements:

Variable 1: the population size measured in number of inhabitants.
Variable 2: the number of public swimming baths in the town.

Since variable 1 is many orders of magnitude greater than variable 2 it is apparent that the former will dominate the result. The distance between town $\mathbf{x} = (x_1, x_2)$ and town $\mathbf{y} = (y_1, y_2)$ is

$$\begin{aligned} d(\mathbf{x}, \mathbf{y}) &= \{(x_1 - y_1)^2 + (x_2 - y_2)^2\}^{1/2} \\ &\simeq (x_1 - y_1) \end{aligned}$$

Variable 2 has been virtually ignored. In such cases it is futile using unmodified variables and we must somehow normalize or scale them so that they are more comparable. A common normalization method is to equalize the variances of the variables (estimated from the classified samples). x_1 is thus redefined to be x_1/s_1, with s_1^2 an estimate of the variance of variable 1. An alternative is to equalize the ranges of the variables, with estimates of the ranges again being obtained from the classified samples.

It might seem that such normalization has resolved the problems of combining different variables, but such is not the case. It has merely replaced one assumption (that variable 1 was much more important than variable 2 in the above example) with another one (that the variables are equally important). Fortunately, some techniques are invariant to such transformations and will give the same result in each case.

1.6 OVERVIEW

In this chapter we have introduced the multivariate space representation for the classification problem and outlined the elements of decision theory. If the prior probabilities, $P(\omega_i)$, and the class-conditional pdfs, $p(\mathbf{x} \mid \omega_i)$ were known, we could stop there. However, as we pointed in Section 1.3, these probabilities are usually unknown and will have to be estimated from samples from the classes. Estimation of the prior probabilities is conceptually straightforward and can be done either from prior theoretical knowledge or on the basis of relative sample sizes. However, estimation of the class-conditional pdfs or, as outlined in Section 1.3, the equivalent estimation of discriminant functions, is more difficult.

Perhaps the most difficult situation occurs when we know nothing at all about the class-conditional pdfs—not even theoretically based knowledge about their parametric form. Chapter 2 outlincs methods which can be applied in such a case, starting with the statement of a rather disappointing theorem due to Rosenblatt (1956) showing that any non-parametric pdf estimate based on a finite sample must be biased.

The most popular non-parametric pdf estimate is, of course, the histogram, and we consider the possibility of using this method here. Unfortunately, the number of histogram cells increases exponentially with increasing dimensionality. This problem, known as Bellman's *curse of dimensionality*, will plague us repeatedly in some form or other.

One way to evade the problem in pdf estimation is to make the cell locations depend on the data, and this is what the kernel estimator, the next subject of Chapter 2, does. We discuss some simple properties of kernel estimators and outline how to choose a good kernel shape and smoothing parameter.

The kernel method basically calculates a probability estimate based on the proportion of sample points falling in a specified volume. The k-nearest-neighbour method (k-NN) does the reverse, specifying the proportion and finding the volume. We introduce this method and its properties and show how it can lead simply into a discriminant function formulation where explicit pdf estimation is avoided.

One of the disadvantages of non-parametric pdf estimation methods is that large data bases must be retained. We outline extensions of the kernel and k-NN methods which allow these stores to be reduced, as well as leading to quicker computation.

For univariate problems a further type of pdf estimator, the series method, has also gained popularity. For completeness we outline the method but point out that it has limited applicability for us, again because of the curse of dimensionality.

Chapter 2 ends with some simple demonstrations of the relationships between the methods and an outline of their relative practical advantages and disadvantages.

It is often the case that we are prepared to make some assumptions about the form of the class-conditional pdfs, for example that they belong to a family of functions indexed by parameters, with only the parameter values being unknown. In this case we can use the samples to estimate the parameters of the class-conditional pdfs. In Chapter 3 we outline desirable properties for such parameter estimates and consider, in turn, maximum likelihood, distance minimization, Bayes methods, and sequential methods of estimation. To illustrate the general results we examine the special cases of normal class-conditional pdfs and, more generally, normal mixture class-conditional pdfs. The normal assumption is a popular one, especially in classical multivariate statistics, and there are sound theoretical reasons for its popularity. However, it is often the case that such an assumption is manifestly inappropriate and yet no alternative is suggested by theoretical considerations. In such cases a normal mixture may be acceptable.

In Section 1.3 we commented that computational advantages could result if discriminant functions were used in place of pdfs. For many problems this is true even to the extent that computational advantages of discriminant functions known to be suboptimal can outweigh their suboptimality disadvantages. Chapter 4 introduces a particular class of functions of this type, namely linear discriminant functions. Because of their analytic tractability there is a large body of literature on this type of discriminant function and in a mere single chapter we can only touch on it. We have, however, outlined four major estimation methods: the linear programming formulation, error correction approaches, Fisher's method, and least squares methods. The various methods have different attributes, and these are discussed.

An approach to linear discriminant functions which has been popular in the statistical literature is via logistic transformations and this approach is outlined.

Separability, i.e. whether two classes of samples can be perfectly separated by a surface of specified form, is of considerable historical significance, as will be apparent from our outline of linear classifiers. There are many ways in which a simple linear discriminant function distinguishing between two classes can be generalized to deal with several classes and we discuss some of these methods. Another generalization, and one which very significantly extends the power of the linear discriminant function, is to redefine the variables in the linear form as being functions of the original measurements.

Up to this point we have only been considering continuous variables. Of equal importance, however, especially in medical and sociological fields, are categorical variables. The ones which have received most attention are the simplest (at the opposite extreme from continuity), namely variables with two categories: dichotomous or binary variables. Reasons for their importance include the advent of the electronic computer, with its emphasis on binary switching, and problems with other categorical variables due to the curse of dimensionality. Explicit estimation of the multinomial distribution functions is in many ways similar to the histograms of Chapter 2, and has the same major problem—that of the curse of dimensionality. All three of the continuous variable non-parametric pdf estimation methods have been extended to binary variables and we outline these extensions, as well as their associated advantages and disadvantages. If the dimensionality is sufficiently low it is also possible to use log-linear models.

Chapter 5 concludes by presenting discriminant function methods for binary data, including the use of linear decision surfaces (after transformation) for the two-class problem and logistic approaches.

The quality of the results of a statistical analysis can be no better than the quality of the data which is subjected to that analysis. This means that great care must be taken in choosing variables and deciding how to measure them. For many problems, and this is particularly true of discrimination and classification, there is a virtually unlimited set of variables from which a choice must be made. Now, while it might seem that the more variables the better on the grounds that adding variables can only improve performance, but never degrade it, this is not quite true. For a fixed size of sample on which to design the decision rule, classification performance can deteriorate with increasing number of variables. This apparent paradox is discussed in Chapter 6 where we consider statistical aspects of variable selection. In that chapter we point out that the division into variable selection and classification rule design is somewhat artificial, but that it has practical advantages. We discuss different types of class separability measure, giving examples of each type and stressing their relative merits for practical variable selection.

It is convenient to divide this problem into two areas: variable *selection*, when the problem is to select a subset of variables; and variable *transformation*, when all of the original measurements can be used. Selection is appropriate if cost or other factors mean that not all of the original set of variables can be measured, and transformation applies when they can all be measured but when increased reliability results if a lower dimensional space is used. The first situation presents an interesting problem in combinatorial analysis (ideally we would like to compare all sets of d' variables which can be chosen from d, and in typical problems facing us $\binom{d}{d'}$ is astronomical). The second situation has been tackled by both linear and non-linear approaches.

Although this book is chiefly concerned with discrimination and classification, i.e. estimating and applying classification rules derived from samples with known classifications, we mentioned in Section 1.1 that there is another similar problem which in a sense logically precedes this one. That is, of course, the problem of cluster analysis, where a sample is obtained but where the correct classifications of the elements of that sample are unknown. Correct classifications might be unknown for several reasons: for example, because no 'correct' classifications exist, because the cost of acquiring a classified set is too great, because the class structure changes with time, etc.

Chapter 7 introduces methods of cluster analysis, pointing out the conceptual similarities to variable selection and showing how variable selection for discrimination differs from that for cluster analysis. In common with other treatments of the subject we have divided the methods into two classes: hierarchical methods (forming clusters by building up from individual points or by breaking down from a single all-inclusive clusters) and optimization methods (allowing reallocation of points to optimize some criterion function).

We again give examples of each and attempt to outline their practical advantages and disadvantages. Methods based on pdf estimation—derived from the mixture methods of Chapter 3—can also be used here. The reader should note, however, that when these methods are applied to *defining* classes, as opposed to merely estimating pdfs, an extra problem can arise. This problem, that of *identifiability*, is outlined and its importance is discussed.

Cluster analysis as a formal area of statistics is a relative new-comer and as such it is faced with the sort of problems which afflict any relatively new area. In Chapter 7 we discuss the particular problems of cluster definition, the need to avoid irrelevant partitions, the problem of normalization prior to a cluster analysis, and how to decide on the number of clusters.

The final chapter, Chapter 8, deals with topics which it was felt did not merit chapters of their own in a book of this size (though clearly each is capable of filling several volumes in its own right). The first of these topics is the assessment of a classifier's performance, which we measure by gauging the misclassification rate. If a limited design set is available this problem is surprisingly difficult and has been the focus of much research. Two particular methods—Lachenbruch's leaving-one-out method and the bootstrap method have very general applicability.

The second problem we consider in Chapter 8 is that of incomplete data: it is a very common occurrence in statistical problems for the data set to have some items missing. We can distinguish two cases here: first, classifying a new, incomplete vector; secondly, estimating a decision surface from incomplete data. Several approaches to the first problem are outlined, including the optimal one of making classifications in relevant subspaces. The second problem is more difficult, but suggestions are made for both the parametric and the non-parametric case.

The third problem is that of incorrectly classified design set elements. This can be viewed as intermediate between supervised pattern recognition, where (for the two class case) the probability that a teacher classification is correct is 1, and unsupervised pattern recognition (cluster analysis) where the probability is $\frac{1}{2}$.

Finally, we consider the important possibility of postponing a classification if one has not sufficient confidence that it is not correct. This deferment possibility is known as the *reject option*.

EXERCISES

1.1 A mechanical seed-sorting device measures the colour of the seeds by optical methods and then, by means of an air jet, deposits the seeds into one of two containers. It is known that there are twice as many type 2 seeds as type 1. Of the type 1 seeds a proportion 0.9 are red, the remainder black. Of the type 2 seeds a proportion 0.2 are red and 0.8 black. Devise a classification rule to minimize the proportion of misclassifications.

If there were five times as many type 2 as type 1 seeds, would your rule be any different?

1.2 In statistical hypothesis testing error probabilities can be reduced by increasing the sample size. Does the same apply in discrimination and classification problems?

1.3 In Section 1.3 it was pointed out that simplified classification rules could often be based on the observation that it was the relative rather than the absolute values of the class-conditional pdfs which mattered. Apply this idea when there are two normal classes with identical variance–covariance matrices.

1.4 Prove that the Euclidean metric is invariant to rotations.

1.5 If one does not have sufficient confidence in a classification it might be advantageous to defer classifying an object. What would be a suitable measure of confidence?

1.6 Defining a two-class classifier by its Ω_1, Ω_2 regions, the minimal error rate is

$$\varepsilon = \int_{\Omega_1} P(\omega_2)p(\mathbf{x}\,|\,\omega_1)\,\mathrm{d}\mathbf{x} + \int_{\Omega_2} P(\omega_1)p(\mathbf{x}\,|\,\omega_1)\,\mathrm{d}\mathbf{x}$$

The Bayes error rate for these probability distributions is

$$\varepsilon^* = \inf_{\Omega_1} \varepsilon$$

Show that ε^* is a concave function of $p(\mathbf{x}\,|\,\omega_i)$. Hence, show that a classification based on the Bayes rule is relatively unaffected by small changes in $p(\mathbf{x}\,|\,\omega_i)$.

CHAPTER 2

Distribution-Free Methods

2.1 INTRODUCTION

If the forms of the class-conditonal pdfs $p(\mathbf{x} | \omega_i)$ are known, then discrimination and classification problems can be solved by comparing the ratios of these functions with various thresholds. Usually, however, these functions will not be known. In some cases we will know the general parametric form of the pdfs (perhaps from theoretical knowledge or by studying the distribution of samples) and in these cases we can use the sample points to give us estimates of their parameters. More generally, we can use the sample distributions to estimate the parameters of discriminant functions or decision surfaces. In many cases, however, we will not be able to make simplifying assumptions about the pdfs or decision surfaces and in such cases we must resort to *distribution-free* or *non-parametric* methods.

There are four major types of non-parametric pdf estimator: the histogram, the kernel method, the k-nearest-neighbour (k-NN) method, and the series method. Each of these methods has different advantages and disadvantages and these will be made apparent in the succeeding sections. We begin with the statement of a somewhat disappointing theorem due to M. Rosenblatt (1956):

Theorem

Let $x_1, \ldots, x_n$ be independent and identically distributed random variables with continuous density function $p(y)$. Let $\hat{p}(y; x_1, \ldots, x_n)$ be an estimate of $p(y)$. The function $\hat{p}(y; x_1, \ldots, x_n)$ is assumed to be jointly Borel measurable in $(y; x_1, \ldots, x_n)$ and it is also assumed that

$$\hat{p}(y; x_1, \ldots, x_n) \geqslant 0$$

since $p(y) \geqslant 0$.

Then $\hat{p}(y; x_1, \ldots, x_n)$ is not an unbiased estimate of $p(y)$. ■

In principle we can make this finite sample bias as small as we like by increasing the sample size. In practice there are, of course, cost and computer store limitations, We end up by compromising by making an implicit assumption about the degree of irregularity in the pdf that our estimate can match. For histograms this *smoothness* depends on the cell size; for kernel estimates it depends on an explicit smoothing parameter; for k-NN estimates it depends on k; and for series estimators it depends on the number of terms in the series.

2.2 HISTOGRAMS

The generalization of the ordinary histogram from one dimension to many is quite straightforward: instead of partitioning the line into disjoint intervals of equal length, we partition the whole space into disjoint cells of equal volume. Then, just as with the univariate case, we estimate the pdf by the proportion of sample points which fall in each cell. This method has the advantage (not shared by the kernel or basic k-NN methods) that the points themselves do not need to be stored after the estimate has been made. Only statistics describing the cell locations, shapes, and sizes need be retained. However, as the number of dimensions increases we encounter a problem which has become known as Bellman's *curse of dimensionality*. With one variable, divided into M intervals there are M cells. With two variables, if each is divided into M intervals, we have M^2 cells. In general, with d variables there are M^d cells. In a practical problem we might have ten intervals for each of eight variables, i.e. one hundred million cells. Since the pdf will be estimated by the proportion of sample points falling in each cell it is clear that either we require an absurdly large number of observations, or the estimate is zero almost everywhere. Not a very satisfactory situation.

Another disadvantage of the histogram, and one which occurs in the univariate case as well as in multivariate cases, is that each cell boundary is a discontinuity—there is a sudden jump from the level of one cell to that of its neighbour. Note also that beyond the boundary cells (the top and bottom ones in one dimension) the estimate falls abruptly to zero, not decaying gradually as does the pdf itself in most cases in practice.

The first of these disadvantages, the prohibitively large number of observations needed in even relatively low dimensional problems, can be alleviated by discarding the idea of a rigid network of cells. If, instead, we let the data somehow determine the cell locations, shapes, and sizes, then we can represent the data with comparatively few cells.

This is the approach suggested by Sebestyen and Edie (1966). For each class their algorithm takes the sample points one at a time and begins by centring the first histogram cell at the first sample point. Each cell is hyper-ellipsoidal in shape with initial dimensions chosen as shown below. Subsequent points may either fall in existing cells, in a *guard zone* about existing cells, or completely outside all cells. In the latter case the point serves as the centre of a new cell. If a point falls in a cell then that cell is updated by having its centre and dimensions recalculated, as shown below. If a point falls in a guard zone then the point is temporarily stored until later in the process when all stored points are used to update their nearest cell. Figure 2.1 illustrates a

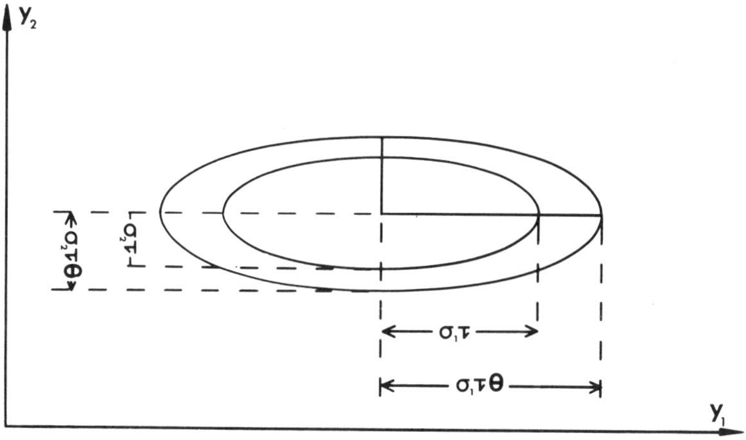

Figure 2.1 A basic cell in the Sebestyen and Edie multivariate histogram method

basic isolated cell in two dimensions, y_1 and y_2. It is not difficult to show that cells tend to migrate towards regions of higher density, thus siting themselves over modes of probability. This means that the method can also be used for cluster analysis (see Chapter 7).

We shall begin the formal description of the algorithm by supposing that the processing has already begun and is at the stage of having c cells $\Gamma_1, \ldots \Gamma_c$ with respective centres $\boldsymbol{\mu}_1, \ldots, \boldsymbol{\mu}_c$. The ith cell will be further described by d variances (d being the number of dimensions) $\sigma_{il}^2, \ldots, \sigma_{id}$ and will currently contain S_i sample points. The cell structure is then updated by modifying the cells to include the next point, $\mathbf{x}$, as follows:

(1) The distance between $\mathbf{x}$ and the centres of each of the cells are calculated using

$$d(\mathbf{x}, \boldsymbol{\mu}_i) = \sum_j (\mathbf{x}_j - \mu_{ij})^2/\sigma_{ij}^2$$

(2) $\boldsymbol{\mu}_k$ is found such that $d(\mathbf{x}, \boldsymbol{\mu}_k) = \min_i d(\mathbf{x}, \boldsymbol{\mu}_i)$
(3) Then, if

$d(\mathbf{x}, \boldsymbol{\mu}_k) \leq \tau$, $\mathbf{x}$ is assigned to cell Γ_k
$\tau < d(\mathbf{x}, \boldsymbol{\mu}_k) \leq \theta\tau$, $\mathbf{x}$ falls in the guard zone of cell Γ_k and is temporarily put aside
$d(\mathbf{x}, \boldsymbol{\mu}_k) > \theta\tau$, $\mathbf{x}$ becomes the centre of a new cell
($\theta(>1)$ and τ are calculated as below).

(4) When a point falls in a cell Γ_k that cell's parameters ($\boldsymbol{\mu}_k$ and $\boldsymbol{\sigma}_k^2$) are updated

$$\boldsymbol{\mu}_k(S_k + 1) = \{\boldsymbol{\mu}_k(S_k) \cdot S_k + \mathbf{x}\}/(S_k + 1)$$

$$\sigma_{ki}^2(S_k+1) = \max\left\{\sigma_i^2(0), \left[\sigma_{ki}^2(S_k) + \frac{(\mu_{ki}(S_k) - x_i)^2}{S_k+1}\right]\frac{S_k}{S_k+1}\right\}$$

where $i = 1, \ldots, d$

S_k is increased by 1
($\sigma_i^2(0)$ is an initial value, described below).

The reader will notice that this updating procedure is merely a sequential way of calculating the means and variances of the points defining each cell. It would be possible to recalculate them from scratch using the standard definitions each time a new point fell in the cell. However, not only would this require more computer time but it would also defeat the object of sequential methods which is that each point can be discarded after it has been processed.

Sebestyen and Edie use the average number of points per cell as a measure of how well the cell structure is established. When the ratio (number of points input)/(number of cells created) reaches a threshold ω the points in the guard zones are used to update the existing cell structure. The temporary storage process then starts again and continues until the ratio (number of points input since last update)/(total number of cells created) reaches 2ω when updating occurs. This is repeated for thresholds $2^2\omega$, $2^3\omega$, etc. When all of the sample points (for the class in question) have been processed any left in temporary storage are used to update their nearest cells. Sebestyen and Edie recommend an ω-value of 4.

One way to find suitable values for τ, θ, and $\sigma_i^2(0)$ is experimentally by trying various different values. Mucciardi and Gose (1972), however, have developed more formal methods and tested them on both real and synthetic data. Their methods are as follows.

(1) $\sigma_i^2(0)$. The marginal distribution of each class for each variable is fitted by a mixture of two normal distributions (see Chapter 3). The larger of the two standard deviations is chosen for each class and each variable (and is equal to $\hat{\sigma}_{ki}$, say). Then, for each variable the *smallest* of these $\hat{\sigma}_{ki}$ is found (σ_i^*, say). $\sigma_i(0)$ is taken as proportional to this smallest value, and is the same for all classes. The constant of proportionality is chosen so that

$$\prod_{i=1}^{d} \sigma_i(0) = 1$$

(2) τ. An effective value for τ can be found by considering a hypersphere centred at a random sample point of the class in question and letting τ be the hypersphere's radius when it contains on average three extra sample points. (Mucciardi and Gose's program chose three random sample points, found the radii of the spheres centred at these points which contained three extra sample points, and set τ to the average of these three radii.)

(3) θ. Mucciardi and Gose found θ by using the same method as for τ, but using spheres containing the five nearest sample points instead of three. The

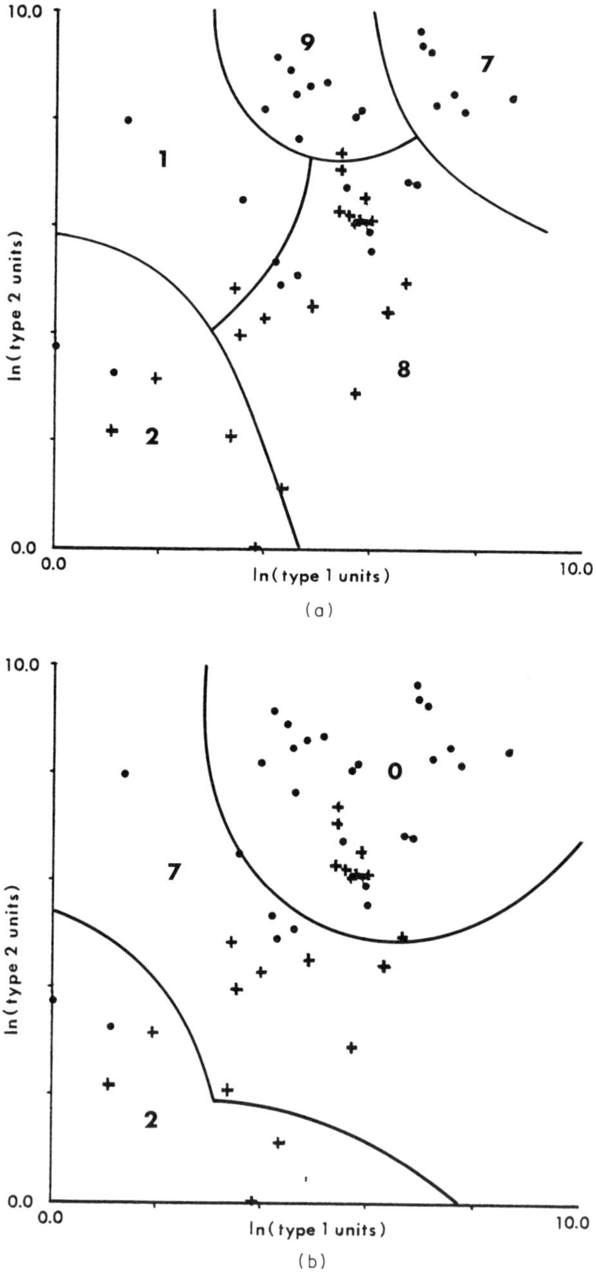

Figure 2.2 (a) Sebestyen and Edie histograms on computer-user data. The numbers indicate the height of the histogram cells. (a) Class 1 histogram (non-medical institutions, dots); (b) class 2 histogram (medical institutions, crosses)

average radius of these spheres gives $\theta\tau$. These authors also suggest that the average squared distance to the kth nearest sample point is approximately proportional to the number of dimensions. If there are many dimensions this linear relationship can save a great deal of computation effort.

We commented above that there were other problems associated with histograms, namely the discontinuities at the edges of the cells and the abrupt drop to zero outside the boundary cells. Sebestyen and Edie overcome the second of these problems by using a local normal decay for the density at the cells' edges (they facilitate implementation by using a step-wise approximation). Mucciardi and Gose overcome both problems by replacing each cell by a normal distribution with a weight proportional to the number of points in that cell.

An illustration of this histogram approach is given in Figures 2.2 and 2.3 applied to the data of Table 2.1. Two variables (amount of computer usage under each of two different operating systems) have been recorded for institutional users of the University of London Computer Centre. The users have been divided into two classes: medical versus non-medical institutions. Because of the skewness of the distributions natural log transformations have been applied to the two variables and for convenience any institution scoring zero has been omitted from the analysis. The parameter values chosen (arbitrarily, not following Mucciardi and Gose) were $\tau = 2.0$, $\theta = 1.5$, $\omega = 4.0$, and $\sigma_i^2(0) = 0.5$ for both classes. For class 1 no forcing of guard zone points into the existing structure occurred until the end of the process,

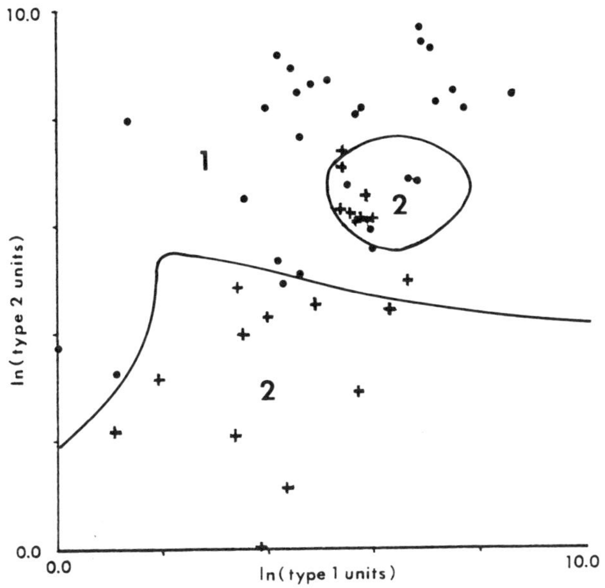

Figure 2.3 Classifier derived from Sebestyen and Edie analysis of computer-user data. Numbers indicate classification regions Ω_i

Table 2.1 Users of the University of London Computer Centre divided into non-medical and medical users. The measurements are the numbers of units of computing used under two different operating systems

Institution	Type 1 units	Type 2 units
Class 1		
Institute of Archaeology	0	0
Institute of Education	379	223
Institute of Historical Research	0	0
Reactor Centre, Silwood Park	74	132
School Examinations Department	1	43
Bedford College	750	943
Birkbeck College	2,177	3,486
Chelsea College	371	370
Imperial College	1,081	10,551
King's College	1,210	3,898
London School of Economics	5,329	4,659
Queen Elizabeth College	1,760	5,012
Queen Mary College	116	5,383
Royal Holloway College	909	11,995
University College London	914	15,625
Westfield College	248	882
Wye College	68	210
Goldsmiths' College	5	0
London Business School	0	7
The University of Reading	99	4,696
University of Surrey	51	3,526
The University of Sussex	0	7
The City University	63	9,963
The University of Southampton	95	2,107
Brunel University	167	6,018
University of Kent at Canterbury	304	3,560
University of Oxford	285	3,383
University College, Cardiff	93	166
University of Bath	6	0
University of Bristol	82	7,782
University of Exeter	4	3,039
UWIST	35	690
The Open University	3	26
Numerical Algorithms Group	0	227
External Users	902	904
Class 2		
School of Pharmacy	134	92
Charing Cross Hospital Med School	142	0
Guy's Hospital Medical School	228	1,187
King's College Hospital Medical School	54	75
The London Hospital Medical College	277	510
The Middlesex Hospital Medical School	219	1,650
Royal Dental Hospital of London	34	54
Royal Free Hospital School of Medicine	297	435
St Bartholomew's Hospital Medical College	404	466

Table 2.1 *continued*

Institution	Type 1 units	Type 2 units
St George's Hospital Medical School	294	18
St Mary's Hospital Medical School	557	85
St Thomas's Hospital Medical School	220	542
University College Hospital Medical School	30	8
Westminster Medical School	31	124
Institute of Basic Medical Sciences	7	24
Institute of Cancer Research	357	712
Cardiothoracic Institute	80	3
Institute of Child Health	321	452
Institute of Laryngology and Otology	0	0
Institute of Neurology	50	0
Institute of Obstetrics and Gynaecology	3	9
Institute of Ophthalmology	19	0
Institute of Psychiatry	374	446
London School of Hygiene	788	142
Royal Postgraduate Medical School	47	1

by which time six cells had been created as listed in Table (2.2(a). For class 2 the guard zone points were forced in after the addition of point 9 (when (number of points)/(number of cells) = $9/2 > \omega = 4$) as well as at the end. The final class 2 cell configuration is shown in Table 2.2(b). Figures 2.2(a) and (b) show the final cell structures. Replacing each cell by a normal distribution with weight given by the number of points joining the cell during processing yielded the classifier of Figure 2.3.

Table 2.2 (a) Sebestyen and Edie histogram structure for non-medical users of ULCC computers. (b) Sebestyen and Edie histogram structure for medical users of ULCC computers

Cell	Number of points	Means		Variances	
		(a)			
1	8	5.48	5.89	0.92	0.59
2	2	0.55	3.51	0.50	0.50
3	7	7.35	8.81	0.50	0.50
4	9	4.76	8.45	0.50	0.50
5	1	1.39	8.02	0.50	0.50
6	1	3.56	6.54	0.50	0.50
		(b)			
1	7	4.46	3.87	1.20	0.86
2	10	5.78	6.30	0.50	0.50
3	2	1.52	2.69	0.50	0.50
4	2	4.12	0.55	0.50	0.50

2.3 KERNEL ESTIMATORS

The earliest development of this type of pdf estimator seems to be due to Parzen (1962), although Rosenblatt (1956) introduce a restricted form. We shall begin, for simplicity, in one dimension, by defining a function $\nu(x \mid \omega_m)$ to be the number of class m sample points with values less than or equal to x. n_m is the total sample size for class ω_m. An obvious estimator for the cumulative distribution of x, $D(x \mid \omega_m)$, is $\hat{D}(x \mid \omega_m)$ given by

$$\hat{D}(x \mid \omega_m) = \frac{\text{number of class } \omega_m \text{ sample points} \leqslant x}{\text{total number of class } \omega_m \text{ sample points}} = \frac{\nu(x \mid \omega_m)}{n_m}$$

Density functions are derivatives of cumulative distribution functions, but we cannot obtain a density function estimator $\hat{p}(x \mid \omega_m)$ simply by differentiating $\hat{D}(x \mid \omega_m)$ because that would merely result in a collection of probability spikes, one at each of the sample points, with zero values everywhere else. However, we can define a density function estimator $\hat{p}(x \mid \omega_m)$ as being an approximation to the derivative of $\hat{D}(x \mid \omega_m)$. Thus

$$\hat{p}(x \mid \omega_m) = \frac{\hat{D}(x + h \mid \omega_m) - \hat{D}(x - h \mid \omega_m)}{2h}.$$

$\hat{p}(x \mid \omega_m)$ defined in this way can also be seen to be the proportion of class ω_m sample points falling in $(x - h, x + h)$ divided by $2h$

$$\hat{p}(x \mid \omega_m) = \frac{\nu(x + h \mid \omega_m)/n_m - \nu(x - h \mid \omega_m)/n_m}{2h}$$

$$= \frac{\nu(x + h \mid \omega_m) - \nu(x - h \mid \omega_m)}{2n_m h} \tag{1}$$

This allows us to rewrite $\hat{p}(x \mid \omega_m)$ as

$$\hat{p}(x \mid \omega_m) = \frac{1}{n_m h} \sum_{i=1}^{n_m} K_0\left(\frac{x - x_i}{h}\right) \tag{2}$$

where x_i, $i = 1, \ldots, n_m$, is the sample and

$$K_0(z) = \begin{cases} 0, & \text{for } |z| > 1 \\ \frac{1}{2}, & \text{for } |z| \leqslant 1 \end{cases} \tag{3}$$

Put in this way it is clear that any point in the interval $(x - h, x + h)$ contributes $1/2n_m h$ to the estimate at x and at any point outside the interval contributes nothing. In effect we are calculating a weighted sums of points where the weight is $1/2n_m h$ for a point in $(x - h, x + h)$ and 0 outside it. Such a harsh weighting, however, is perhaps rather artificial—why should a point just inside the interval contribute the same as a point very near to x, while a point just outside the interval contributes nothing? We can generalize the estimator by using a smoother weighting function. We could, for instance, let

$$K_0(z) = \frac{1}{\sqrt{2\pi}} \exp\{-\tfrac{1}{2}z^2\} \tag{4}$$

so that *all* the class ω_m sample points, x_i, contribute to the estimate at x, but inversely according to their distance from x. Figure 2.4 shows estimates based

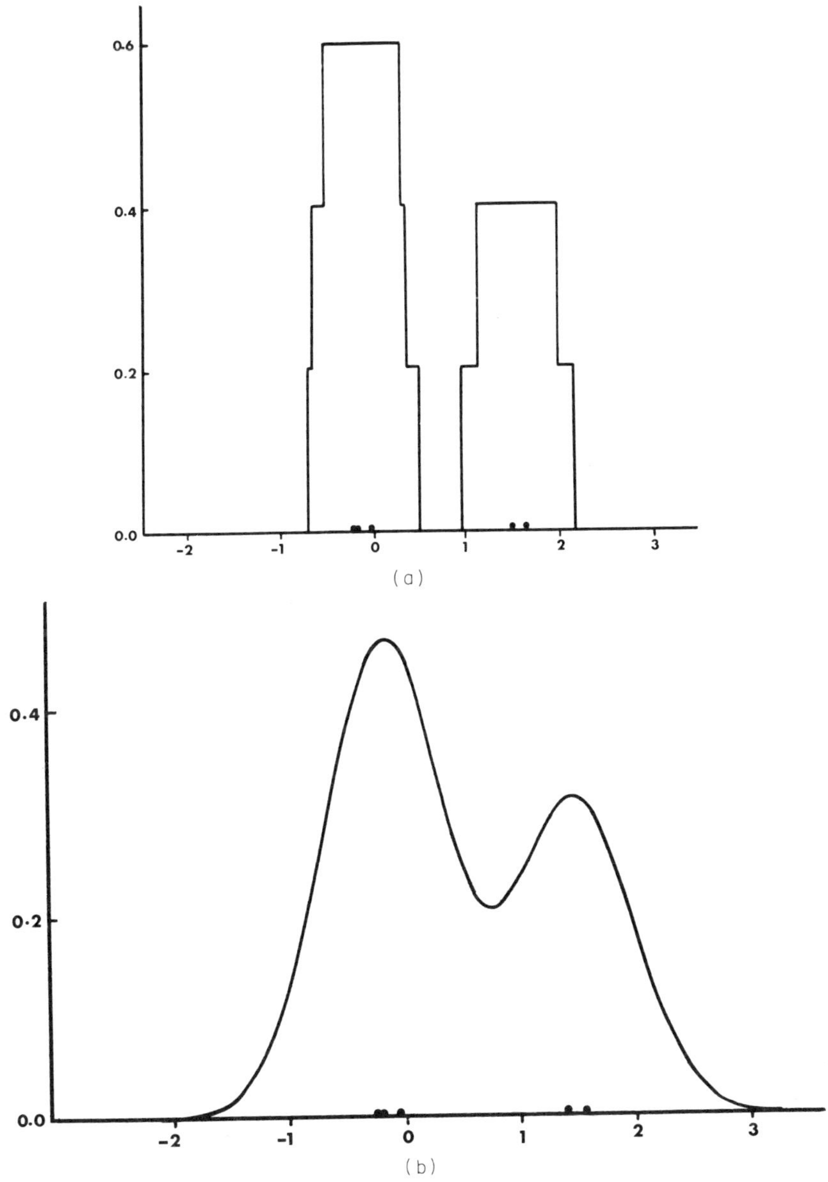

Figure 2.4 (a) Kernel estimate based on a random sample of size 5 from a univariate normal distribution. Here a rectangular kernel with $h = 0.5$ is used. (b) Kernel estimate based on a random sample of size 5 from a univariate normal distribution. Here a normal kernel with $h = 0.5$ is used

on a random sample of size 5 from a standard normal population. Figure 2.4(a) shows the estimate using kernel (3) above and 2.4(b) the estimate using (4). Both have $h = 0.5$. Readers familiar with time series analysis will recognize the similarity between this estimator and moving averages. K_0 is called the *kernel* function.

Extension of (2) to higher dimensions is straightforward. For example, for multivariate normal kernels (the extension of (4)) we have (see, for example, Shanmugan, 1977)

$$\hat{p}(\mathbf{x}|\omega_m) = \frac{1}{n_m}\sum_{i=1}^{n_m} \frac{1}{(2\pi)^{(d/2)}|\mathbf{\Sigma}|^{1/2}} \exp[-\tfrac{1}{2}(\mathbf{x}-\mathbf{x}_i)'\mathbf{\Sigma}^{-1}(\mathbf{x}-\mathbf{x}_i)]$$

Here $\mathbf{x}$ is the point at which the estimate is being made and $\mathbf{x}_i$, $i = 1, \ldots, n_m$, is the design sample for class ω_m. Frequently the amount of calculation involved in this estimate is excessive and $\mathbf{\Sigma}$ is replaced by a diagonal matrix to give

$$\hat{p}(\mathbf{x}|\omega_m) = \frac{1}{n_m}\sum_{i=1}^{n_m} \frac{1}{h_1 \ldots h_d(2\pi)^{(d/2)}} \prod_{j=1}^{d} \exp\left[-\frac{1}{2}\frac{(x_j - x_{ij})^2}{h_j}\right]$$

Often even more simplification is made, setting $h_j = h$ for $j = 1, \ldots, d$, so that

$$p(\mathbf{x}|\omega_m) = \frac{1}{n_m h^d (2\pi)^{(d/2)}} \sum_{i=1}^{n_m} \prod_{j=1}^{d} \exp\left[-\frac{1}{2}\left(\frac{x_j - x_{ij}}{h}\right)^2\right]$$

For obvious reasons such a multivariate kernel is called a *product kernel*. Although we have used the normal kernel as an example, other shapes can easily be used. A general product kernel estimator with equal h-values is

$$\hat{p}(\mathbf{x}|\omega_m) = \frac{1}{n_m h^d} \sum_{i=1}^{n_m} \prod_{j=1}^{d} K_0\left(\frac{x_j - x_{ij}}{h}\right)$$

Note that in fact this allows the possibility of having quite different K_0 functions in different dimensions.

Whether or not a product kernel is being used, we can write the general form of kernel estimator as

$$\hat{p}(\mathbf{x}|\omega_m) = \frac{1}{n_m}\sum_{i=1}^{n_m} K(\mathbf{x} - \mathbf{x}_i)$$

If we impose the conditions $K(\mathbf{z}) \geq 0$ and $\int K(\mathbf{z})\, d\mathbf{z} = 1$ on K, then it is not difficult to see that $\hat{p}$ also satisfies $\hat{p}(\mathbf{z}) \geq 0$ and $\int \hat{p}(\mathbf{z})\, d\mathbf{z} = 1$ so that $\hat{p}(\mathbf{z})$ is a legitimate pdf.

The role of h in these estimators is apparent from the particular one-dimensional examples illustrated in Figure 2.4. For a fixed kernel shape, h determines how much each sample point contributes to the estimate at any point, x. h is thus a *spread* or *smoothing* parameter. (In fact we can rewrite (2) in the form

$$\hat{p}(x \mid \omega_m) = \int \frac{1}{h} K_0\left(\frac{x-y}{h}\right) \mathrm{d}\hat{D}(y \mid \omega_m)$$

whence it becomes apparent that this estimator is no more than a convolution smoothing of the sample points). If h is very small the estimator degenerates into a collection of n_m sharp peaks, each located at a sample point. Conversely, if h is too large, the estimate is oversmoothed and an almost uniform pdf results.

From another viewpoint we can consider smoothing as an attempt to find an acceptable compromise between bias and random fluctuation. This is most easily seen with the rectangular kernel (3) and equation (1). For large h, $\hat{p}(x \mid \omega_m)$ will have small variance since more points will fall in $(x - h, x + h)$. However, for large h a smoother $\hat{p}(x \mid \omega_m)$ results—which yields more bias. Conversely, small h allows $\hat{p})x \mid \omega_m)$ to follow the curvature of $p(x \mid \omega_m)$ more closely but also allows $\hat{p}(x \mid \omega_m)$ to fluctuate more violently.

It is also apparent that h must depend on n_m, the sample size. With very few sample points more smoothing, i.e. a large h, is needed so that separate peaks do not result. With a very large sample, however, a small h can be used without the danger of obtaining separate peaks. We thus require h to be a function of n_m satisfying

$$\lim_{n_m \to \infty} h(n_m) = 0$$

Now, although we know from Section 2.1 that $\hat{p}(x \mid \omega_m)$ will be biased for any finite sample, it is possible to impose conditions on K which, taken in conjunction with the conditions on K and h above, ensure asymptotic unbiasedness. These are

$$\int |K(z)| \,\mathrm{d}z < \infty$$

$$\sup_{s} |K(z)| < \infty$$

$$\lim_{z \to \infty} |zK(z)| = 0$$

If we further require that h satisfies

$$\lim_{n_m \to \infty} n_m h\ (n_m) = \infty$$

then $\hat{p}(x \mid \omega_m)$ is asymptotically consistent. Other properties of $\hat{p}(x \mid \omega_m)$ under various conditions on K and h may be found in Parzen (1962) and Cacoullos (1966).

Although asymptotic properties are interesting, in practice we will be faced with a finite and possibly small) sample. Instead of knowing how h should converge as n_m increases, we merely want to know a good choice of h for a particular n_m. A considerable amount of effort has been expended on this problem and we can make the following observations. First, the optimal h will depend on the criterion used (e.g. mean square error between $p\ (x \mid \omega_m)$ and

$\hat{p}(x \mid \omega_m)$, integrated mean square error, relative global error, etc.). Secondly, an h which is optimal for one class may not be so for another: in simulations using two normal populations with different variance–covariance matrices Van Ness (1979) found a marked advantage in estimating h separately for each population. Thirdly, the optimal h will depend on the kernel used. Finally, what is optimal in one part of the space might not be optimal in others. This fourth point has led Wagner (1975), Habbema *et al.* (1978), Raatgever and Duin (1978), Breiman *et al.* (1977), and others to investigate h-values which are functions of x as well as of n_m. Thus, in regions where the sample is densely concentrated smaller h-values can be used.

In view of these observations, we recommend the following approaches if constant h is to be used.

(a) Plot estimates (or marginal estimates in multivariate cases) for several h-values and choose one by eye which is neither too smooth nor too irregular.

(b) Find the average distance between sample points and their qth nearest neighbour and use this as h. If, for example, we are using normal kernels, this ensures that on average q points lie within one standard deviation of x. It has been suggested that $q = 10$ points is a reasonable value.

(c) Bearing in mind that our ultimate aim is to classify new observations, a good scheme is to try to range of h-values and to assess the misclassification rates of the resulting classifiers (by applying them to an independent sample, but see Chapter 8).

(d) One could try a maximum likelihood approach, finding that h which maximized

$$L_m(h) = \prod_{i=1}^{n} \hat{p}(x_i \mid \omega_m; x_1, \ldots, x_n)$$

but this is maximized by $h = 0$—that is, by an estimate which consists of a probability spike at each x_i and zero probability density elsewhere. To get around this, Habbema *et al.* (1974) (see also Aitchison and Aitken, 1976) consider a 'leaving-one-out' method, replacing the above $L_m(h)$ by

$$L_m(h) = \prod_{i=1}^{n} \hat{p}(x_i \mid \omega_m; x_1, \ldots, x_{i-1}, x_{i+1}, \ldots, x_n)$$

The problem of choosing the kernel shape is just as difficult as that of choosing h. For analytic purposes continuity of $\hat{p}$ and its derivatives can be useful—and normal kernels are popular. On the other hand, other shapes, such as K_0 in (3), are much easier to calculate. Piecewise linear kernels (see Figure 2.5) can be an acceptable compromise. The fact that normal kernels have an unbounded region of support ($K_0(x) > 0$ for all x) means that all sample points contribute to the estimate $\hat{p}(x \mid \omega_m)$ at every x. Kernels with finite regions of support (those for which $K(x) = 0$ for large enough x) lead to quicker computation since K does not need to be evaluated for all the sample points. However, this means that the estimate $\hat{p}(x \mid \omega_m)$ can take zero values,

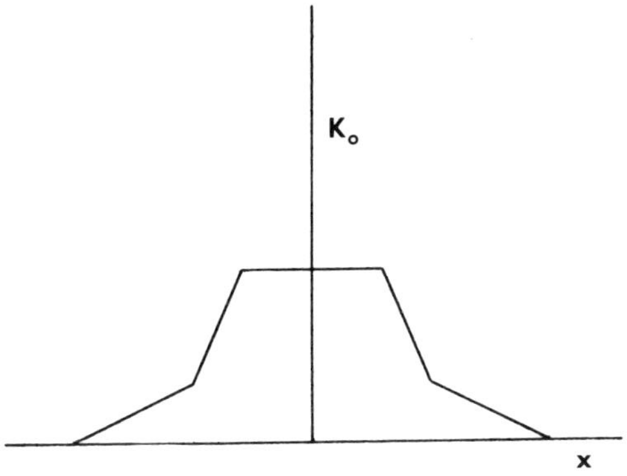

Figure 2.5 A piecewise linear kernel

and this is often a disadvantage, especially in high dimensional spaces. To conclude the discussion of kernel shapes, note that in general any pdf can serve as a kernel—these necessarily satisfy $K_0(x) \geq 0$ and $\int K_0(x)\,dx = 1$.

In general, kernel estimators have attractive mathematical properties and can be usefully applied in practice.

An example of the basic kernel method is given in Figure 2.6 applied to the data of Table 2.3. This data set is from Laurie (1979) and is here used for

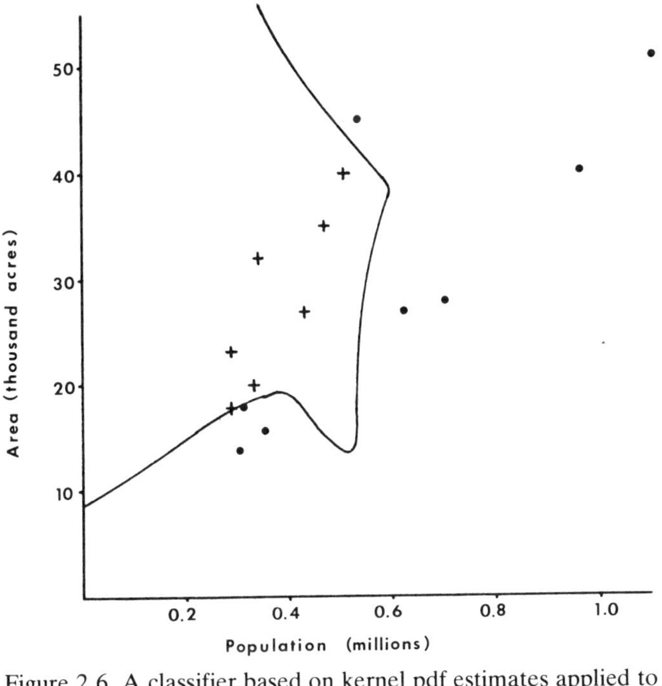

Figure 2.6 A classifier based on kernel pdf estimates applied to the data of Table 2.3

Table 2.3 Estimated number of casualties caused by a nuclear strike on the fifteen largest British cities (London excluded)

City	Population (millions)	Area (thousand acres)	Casualties (%)
Class 1			
Birmingham	1.1	51	85
Manchester	0.62	27	93
Sheffield	0.53	45	89
Liverpool	0.70	28	75
Hull	0.30	14	74
Nottingham	0.31	18	95
Newcastle upon Tyne and Gateshead	0.35	16	71
Glasgow	0.96	40	74
Class 2			
Leeds	0.51	40	62
Leicester	0.29	18	66
Coventry	0.33	20	62
Stoke-on-Trent and Newcastle under Lyme	0.34	32	45
Bristol	0.43	27	50
Cardiff	0.29	23	56
Edinburgh	0.47	35	41

(Reproduced from *Beneath the city streets* by P. Laurie (1979), published by Granada Publishing, with permission from the author)

illustrative purposes only. It consists of two measurements (population and area) on each of the fifteen largest British cities (London omitted). These fifteen cities are divided into two classes, those with an estimated fatality rate of 70 per cent or greater resulting from a nuclear strike, and those with an estimated rate of less than 70 per cent. The decision surface shown is the result of comparing weighted kernel pdf estimates for the two classes. In each case the kernel was bivariate normal with a variance–covariance matrix

$$\mathbf{\Sigma} = \begin{bmatrix} 3.0 & 0.0 \\ 0.0 & 3.0 \end{bmatrix}$$

In Section 2.4 we explain at some length a modification of the nearest neighbour pdf estimation method which leads to reduced computer storage requirements. The same method can be applied here. Briefly, the method is as follows. The data points are processed sequentially and each point is either retained or discarded according to whether it is, respectively, incorrectly or correctly classified by those points retained so far. When all of the points have been processed those rejected are again considered, and included if they are incorrectly classified. This is repeated until there is no change in the set of points retained. The result of this is that one ends up with a design set which is no longer a random sample from the class-conditional pdfs, and thus absolute values of the pdfs cannot be estimated. However, for classification purposes absolute values are not important, it is the relative values which matter (which pdf is greater at a particular point). In practical terms more points are retained near the decision surfaces and fewer are retained in regions far from the decision surface where there is a little doubt as to the correct classification. This approach is sometimes termed the *potential function method*, where an analogy is made between the kernels centred at the data points and potential functions (e.g. electrostatic or gravitational forces) centred at these points.

2.4 *k*-NEAREST-NEIGHBOUR METHODS

For the multivariate pdf $p(\mathbf{x}|\omega_m)$ for class ω_m the probability that a point will fall in a local neighbourhood L of $\mathbf{x}$ is

$$\theta = \int_L p(\mathbf{y}|\omega_m)\,\mathrm{d}\mathbf{y}$$

If L is small and of volume V we can make the approximation

$$\theta \simeq p(\mathbf{x}|\omega_m) \cdot V$$

yielding

$$p(\mathbf{x}|\omega_m) \simeq \theta/V$$

θ/V is, of course, a *smoothed* approximation to $p(\mathbf{x}|\omega_m)$. It is the average value of $p(\mathbf{y}|\omega_m)$ in the local region L around $\mathbf{x}$, and clearly if L is small enough $p(\mathbf{y}|\omega_m)$ in L will not deviate much from $p(\mathbf{x}|\omega_m)$ so that the approximation will be a good one. From this approximation we can develop a pdf estimator: θ can be estimated simply by seeing what proportion of the n_m sample points fall in L. If k is the number of sample points falling in L we have

$$\hat{\theta} = k/n_m$$

which leads to the estimator $\hat{p}(\mathbf{x}|\omega_m)$ defined by

$$\hat{p}(\mathbf{x}|\omega_m) = \frac{k}{n_m V} \tag{1}$$

There are two ways to proceed from this point. First, we could fix the volume V and see how many points, k, fall in this volume. Consider a one-dimensional example. V now becomes the length of an interval—say an interval $(x - h, x + h)$ of length $2h$. Comparing (1) above with (1) of Section 2.3 we see immediately that this is another approach to the kernel estimator. We commented in the preceding section that one of the disadvantages of the ordinary kernel method was that the spread parameter h was the same all over the space. In regions of low probability this could lead to a multiply-peaked estimate, while in regions of high probability the estimate might not be able to match the fluctuations of the pdf. The second way to proceed from (1) avoids this problem in a very natural way by letting V depend on the data: we fix k and determine the volume V needed to enclose the k nearest points to $\mathbf{x}$.

Let us see how k should differ for different n_m. The estimate, k/n_mV, will be subject to less random variation if we take k as large as possible: so k should increase with increasing n_m. Conversely, the errors introduced by the averaging effect will be smaller the smaller V is, and V can be kept small by letting $k/n_m \to 0$ as n_m increases. In fact the two conditions

$$\lim_{n_m \to \infty} k = \infty$$

and

$$\lim_{n_m \to \infty} k/n_m = 0$$

are necessary and sufficient for $\hat{p}(\mathbf{x}|\omega_m)$ to converge in probability to $p(\mathbf{x}|\omega_m)$ at all points of continuity of $p(\mathbf{x}|\omega_m)$. This is a convenient place to comment on a theoretical disadvantage of the k-NN pdf estimator compared with the kernel estimator. This is simply that *the k-NN estimator is not a pdf.* In fact, if (1) is integrated over the whole space (try it for one dimension) we find the integral is not unity but is infinity. Modified estimators have been developed which satisfy the unit integral condition but we have yet to see them applied in practice.

Bearing in mind that our aim is to use the estimates to classify new observation points, let us extend the k-NN methods slightly. Combine all the classes' sample points into one set of n points (so that $\Sigma_m n_m = n$). Now find the hypersphere of volume V which just encloses k points from the combined set. Suppose that amongst these k there occur k_m from class ω_m. Then we can define a k-NN estimator for class ω_m by

$$\hat{p}(\mathbf{x}|\omega_m) = \frac{k_m}{n_m V}$$

(which is slightly different from (1) because V is different). We also have obvious estimators

$$\hat{P}(\omega_m) = \frac{n_m}{n}$$

$$\hat{p}(\mathbf{x}) = \frac{k}{nV}$$

And by Bayes rule

$$\hat{P}(\omega_m|\mathbf{x}) = \frac{\hat{p}(\mathbf{x}|\omega_m)\hat{P}(\omega_m)}{\hat{p}(\mathbf{x})}$$

$$= \left\{\frac{k_m}{n_m V} \cdot \frac{n_m}{n}\right\} \Big/ \left\{\frac{k}{nV}\right\}$$

$$= \frac{k_m}{k}$$

This leads immediately to the classification rule: classify **x** as belonging to class i if $k_i = \max_m(k_m)$. Not surprisingly, known as the *k-nearest-neighbour classification rule*. For the special case when $k = 1$, this is simply termed the nearest-neighbour (NN) classification rule. In practical terms the NN rule classifies **x** as belonging to the same class as that of its nearest neighbour in the entire set of samples. It may happen that, owing to some sampling distortion, the n_m are not proportional to the $P(\omega_m)$. It is not difficult to see that the classification rule then becomes: $\mathbf{x} \in \Omega_i$ if $r_i k_i = \max_m(r_m k_m)$, where $r_m n_m \propto P(\omega_m)$. Brown and Koplowitz (1979) have studied the simple nearest-neighbour rule when $P(\omega_m)$ is not proportional to n_m/n. Although the nearest-neighbour rule will usually lead to a misclassification rate greater than the Bayes rate (the minimum possible), it has the interesting property that its asymptotic misclassification rate is bounded by twice the Bayes rate. We shall prove this for the nearest-neighbour multiclass case below, but before doing so let us just note that the asymptotic misclassification rates for k-NN methods with $k > 1$ have tighter bounds. In general, for k odd it is possible to show that the asymptotic two-class error rate is bounded above by the smallest concave function of ρ^* (the Bayes rate) greater than

$$\sum_{i=0}^{(k-1)/2} \binom{k}{i} [\rho^{*i+1}(1-\rho^*)^{k-i} + \rho^{*k-i}(1-\rho^*)^{i+1}]$$

Returning to the NN case, we have:

Theorem

The N class NN asymptotic misclassification rate is bounded above by twice the Bayes misclassification rate, ρ^*. In fact, we prove a tighter bound, namely

$$r \leqslant \rho^*\left(2 - \frac{N}{N-1}\rho^*\right)$$

Proof. The NN conditional misclassification rate, given **x** and its nearest neighbour $\hat{\mathbf{x}}$, is $r(\mathbf{x}, \hat{\mathbf{x}})$ given by

$$r(\mathbf{x}, \hat{\mathbf{x}}) = P(\omega_1|\mathbf{x})[1 - P(\omega_1|\hat{\mathbf{x}})] + \ldots + P(\omega_N|\mathbf{x})[1 - P(\omega_N|\hat{\mathbf{x}})]$$

$$= 1 - \sum_{i=1}^{N} P(\omega_i|\mathbf{x})P(\omega_i|\hat{\mathbf{x}})$$

Now, as n, the total number of sample points, tends to infinity, so $\hat{\mathbf{x}} \to \mathbf{x}$ and

$$r(\mathbf{x}) \triangleq \lim_{n \to \infty} r(\mathbf{x}, \hat{\mathbf{x}}) = 1 - \sum_{i=1}^{N} P^2(\omega_i | \mathbf{x}) \tag{2}$$

The conditional Bayes misclassification rate $r^*(\mathbf{x})$ is

$$r^*(\mathbf{x}) = 1 - P(\omega_m | \mathbf{x})$$

where $P(\omega_m | \mathbf{x}) = \max_i P(\omega_i | \mathbf{x})$. This allows us to rewrite (2) as

$$r(\mathbf{x}) = 1 - P^2(\omega_m | \mathbf{x}) - \sum_{\substack{i=1 \\ i \neq m}}^{N} P^2(\omega_i | \mathbf{x})$$

$$= 1 - [1 - r^*(\mathbf{x})]^2 - \sum_{\substack{i=1 \\ i \neq m}}^{N} P^2(\omega_i | \mathbf{x}) \tag{3}$$

By the Schwartz inequality we have

$$(N - 1) \sum_{i \neq m} P^2(\omega_i | \mathbf{x}) \geq \left[\sum_{i \neq m} P(\omega_i | \mathbf{x}) \right]^2$$

and using the fact that $\sum_{i=1}^{N} P(\omega_i | \mathbf{x}) = 1$ gives

$$(N - 1) \sum_{i \neq m} P^2(\omega_i | \mathbf{x}) \geq [1 - P(\omega_m | \mathbf{x})]^2 = [r^*(\mathbf{x})]^2$$

Using this in (3) yields

$$r(\mathbf{x}) \leq 1 - [1 - r^*(\mathbf{x})]^2 - \frac{[r^*(\mathbf{x})]^2}{N - 1}$$

i.e.

$$r(\mathbf{x}) \leq 2r^*(\mathbf{x}) - \frac{N}{N-1} \cdot [r^*(\mathbf{x})]^2 \tag{4}$$

Now

$$\text{var}[r^*(\mathbf{x})] = E[r^*(\mathbf{x})^2] - E[r^*(\mathbf{x})]^2 \geq 0$$

so

$$E[r^*(\mathbf{x})^2] \geq E[r^*(\mathbf{x})]^2 = \rho^{*2}$$

and using this in the expectation of (4) gives

$$r \leq \rho^* \left(2 - \frac{N}{N-1} \rho^* \right) \quad \blacksquare$$

If we knew ρ^* then this bound would give us an indication of the worst performance that we could expect from the NN method for large samples. A bound of more practical use is obtained by inverting the above to give

$$\rho^* \geqslant \frac{N-1}{N} - \sqrt{\frac{N-1}{N}}\sqrt{\frac{N-1}{N} - r}$$

which tells us that for large samples the Bayes error rate is above a certain value. That simply means that any classifier based on the chosen variables will always have a misclassification rate higher than this value. We already know, of course, that $\rho^* \leqslant r$, so that a satisfactorily low level of r tells us that the classes are adequately distinct in the space of the chosen variables. Other error bounds on ρ^* arising from nearest-neighbour rules are given by Devijver (1979) who makes use of the reject option. This is discussed in Chapter 8.

A disadvantage common to both the basic kernel and basic k-NN methods is that all of the sample points need to be retained. Not only does this mean that a large part of the computer store has to be set aside for the sample, but it also means that classification can be slow—the distances from $\mathbf{x}$ to all of the sample points must be determined—and this can be important in some applications. However, since all we need to know is the rank order of weighted class-conditional pdfs at $\mathbf{x}$, and not their absolute values, it might be possible to discard some sample points without affecting the rank orders. One way to do this was briefly outlined in the preceding section. Here we present a more detailed explanation with particular reference to nearest-neighbour methods. We begin with Hart's (1968) *condensed-nearest-neighbour* (CNN) rule, which may be described as follows.

There are two areas of store, call them A and B. Initially B contains all the sample points. One point is transferred from B to A to start the process. Subsequently each point in turn is taken from B and compared with those in A. If nearest-neighbour classification of the point with respect to those in A is correct, then that point is replaced in B. If it is incorrect, the point is transferred to A. When all the points of B have been considered, the process is repeated, scanning through the now depleted B set. The process stops when a scan through B fails to transfer any points.

A practical example of the CNN method is given in Figure 2.7 where samples of 30 points drawn from two overlapping circular uniform distributions have been subjected to the CNN algorithm (in fact we used a 5-NN algorithm to classify each point in B with respect to those in A). As can be seen, a sample size reduction of about 50 per cent has occurred.

Of course, once a point has been transferred from B to A there is no possibility of removing it—even though subsequent additions to A might classify it correctly so there is no longer any need to retain it. Gates (1972) devised the *reduced-nearest-neighbour* rule to make removal possible. In brief this algorithm waits until the CNN process is complete and then removes each point in turn from A and tests B on the remaining points. If the new A set correctly classifies B and the 'removed' point, then this point is transferred back to B. Otherwise it remains in A. This fairly obvious extension of the CNN rule leads to a marginally decreased store requirement but it is doubtful that the decrease justifies the large number of extra iterations needed.

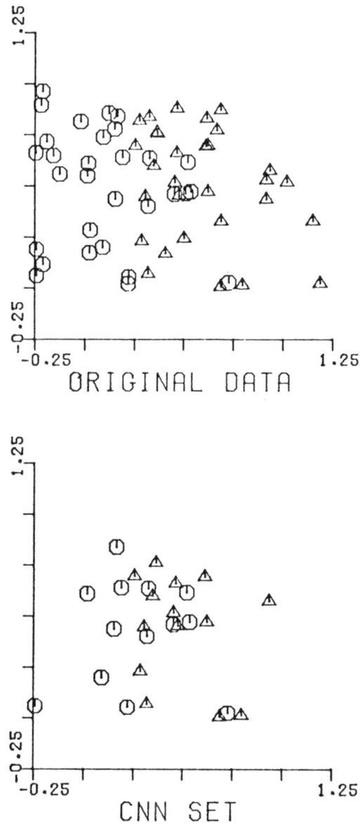

Figure 2.7 The condensed nearest-neighbour method applied to samples of 30 points drawn from two overlapping circular uniform distributions

One of the disadvantages of the straightforward CNN method is that a few outlying points from one class often account for a disproportionately large fraction of the number of points retained from the other class. This is because the CNN method retains outliers from one class which lie in regions densely populated by the other class (since otherwise these outliers would certainly be misclassified). Retention of the outliers means in turn that points from the denser class lying near these outliers must also be retained to prevent them being misclassified. If we could eliminate the outliers then a large number of other points become unnecessary. In effect we would be getting nearer to the Bayes rule since isolated outliers from one class in the midst of large numbers of points from the other class correspond to low probability densities from one class and high densities from the other. Generalizing to more than two classes, we would reject sample points $\mathbf{x} \in \omega_m$ if

$$\frac{p(\mathbf{x} \mid \omega_m)}{p(\mathbf{x} \mid \omega_j)} < t, \quad \text{for some } j$$

where t is a threshold which depends on the prior probabilities and costs (see Section 1.3). Once the whole sample has been processed those points satisfying the above inequality are discarded and a usual CNN analysis is carried out on the remainder; Hand and Batchelor (1978) used a k-NN method for this preprocessing and called the combination of preprocessing and CNN the *edited-nearest-neighbour* rule. The physical effect is to smooth out the decision surfaces, with the degree of smoothing depending on k. Figure 2.8 illustrates the method applied to the same data as Figure 2.7. It is obvious that a very significant store reduction can occur.

Further extensions of the basic CNN rule have been made by other authors (see, for example, Chidananda Gowda, and Krishna, 1979) though usually at the expense of significant extra computer time requirements.

With kernel methods it is necessary to choose a value for h, and similarly with k-NN methods k must be chosen. Loftsgaarden and Quesenberry (1965) suggest that, if using k-NN *estimation*, then for class ω_m a k-value near $\sqrt{n_m}$ is reasonable. Clearly, however, k should depend both on the number of variables and on the smoothness of the pdfs involved. Once again, whether

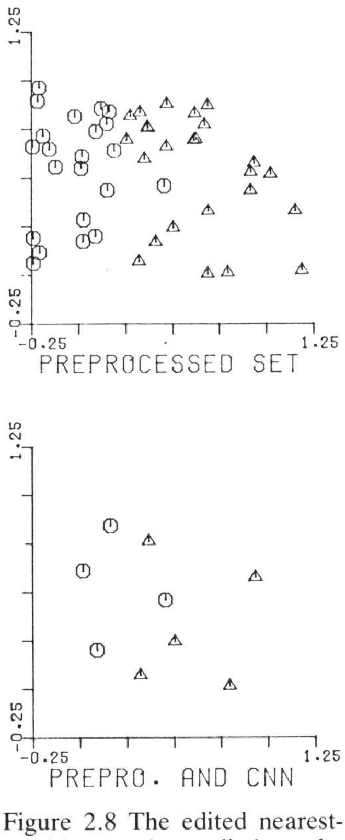

Figure 2.8 The edited nearest-neighbour rule applied to the data of Figure 2.7

using classification via k-NN estimation or direct k-NN classification, a practical approach seems to be trial and error by trying various different k-values and assessing the misclassification rate (see Chapter 8).

For both kernel and k-NN methods as presented above it is necessary to compute the distances from $\mathbf{x}$ to all sample points. Reducing the number of sample points is one way to reduce the amount of computation. Another way (which can be used in conjunction with the above) is the following branch and bound algorithm suggested by Fukunaga and Narendra (1975) for the k-NN method. (The branch and bound method is a combinatorial optimization algorithm which has been applied in many areas of statistics—see, for example, Hand, 1981. As well as in the present application it is used in this book in Chapters 6 and 7.)

The method begins by preprocessing the sample points to impose an 'order' so that not all of the distance calculations need to be made. The basic idea is that the sample set is decomposed hierarchically into disjoint subsets which can then be rejected (or not) as not possibly containing the nearest neighbour to $\mathbf{x}$ without testing each element of the subsets. For simplicity we will illustrate the method with $k = 1$, though extension to $k > 1$ is straightforward.

Step 1: Decomposition

A clustering algorithm (Fukunaga and Narendra use the k-means method—see Chapter 7) is used to split the data into s subsets; then each subset is

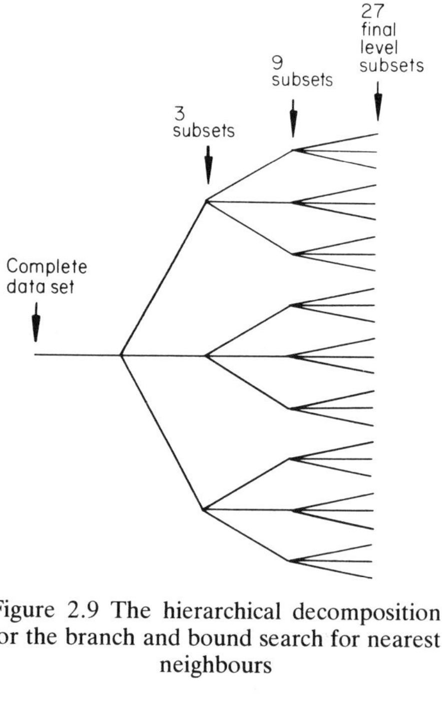

Figure 2.9 The hierarchical decomposition for the branch and bound search for nearest neighbours

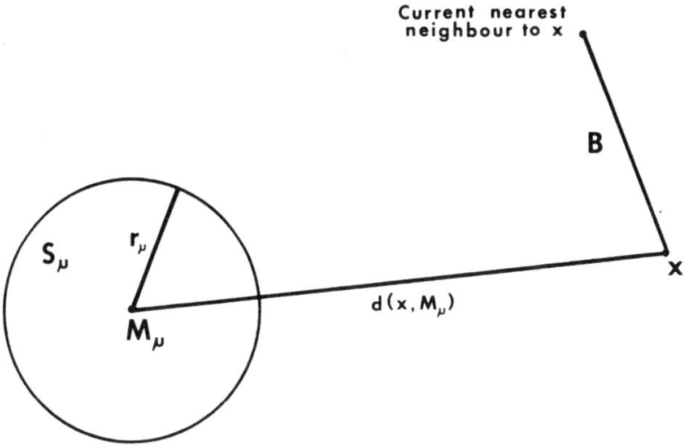

Figure 2.10 Eliminating points in set S_μ as candidate nearest neighbours to **x** by means of the triangle inequality

split into a further s subsets and so on. Figure 2.9 illustrates the resulting decomposition.

Step 2: Tree search

(i) Each node, μ, of the tree is tested to see if the nearest neighbour to **x** can be in the subset, S_μ, associated with μ; if it cannot, then that whole subset is rejected. The testing procedure is illustrated in Figure 2.10 and is as follows.

No $\mathbf{x}_i \in S_\mu$ can be the nearest neighbour to **x** if

$$B + r_\mu < d(\mathbf{x}, \mathbf{M}_\mu)$$

where B is the distance between **x** and its current nearest neighbour, $\mathbf{M}_\mu$ is the mean of S_μ, $d(\mathbf{a}, \mathbf{b})$ is the distance between **a** and **b**, and

$$r_\mu = \max_{\mathbf{x}_i \in S_\mu} d(\mathbf{x}_i, \mathbf{M}_\mu)$$

(ii) For a node, μ, at the final level of the tree, if the node has not been eliminated by the above, then the distances from **x** to the individual samples

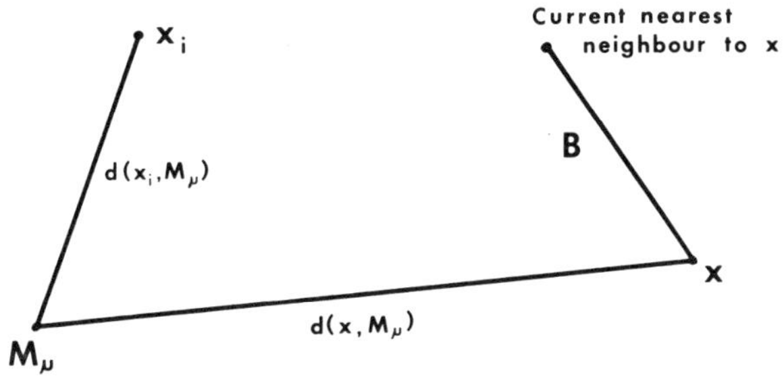

Figure 2.11 There is no need to calculate $d(\mathbf{x}_i, \mathbf{x})$ if $B + d(\mathbf{x}_i, \mathbf{M}_\mu) < d(\mathbf{x}, \mathbf{M}_\mu)$

in S_μ must be calculated. Even so, however, many distance calculations can be avoided by applying another rejection criterion (Figure 2.11). Here $\mathbf{x}_i \in S_\mu$ cannot be the nearest neighbour to $\mathbf{x}$ if

$$B + d(\mathbf{x}_i, \mathbf{M}_\mu) < d(\mathbf{x}, \mathbf{M}_\mu)$$

2.5 SERIES EXPANSIONS

A popular method for approximating waveforms is by expansion in terms of a series of orthonormal basis functions (for example, Fourier series, Legendre polynomials, Hermite polynomials, etc.). Since a one-dimensional pdf can be regarded as a waveform, this is an approach worth considering and Čencov (1962) seems to have been the first to investigate it. Let $p(x \mid \omega_m)$ be the pdf we wish to estimate and let $\{\phi_i\}$ be the set of orthonormal basis functions so that

$$\int \phi_i(x)\phi_j(x)\,\mathrm{d}x = \delta_{ij}$$

The projection of $p(x \mid \omega_m)$ into the space spanned by $\{\phi_i\}$ will be denoted by $p^*(x \mid \omega_m)$ so we have

$$p^*(x \mid \omega_m) = \sum_{i=1}^{\infty} a_i \phi_i(x)$$

where the a_i are coefficients. To estimate $p^*(x \mid \omega_m)$ we must first truncate the series to a finite number, s, of terms, and secondly we must estimate the a_i (by $\hat{a}_i$ say). The $\hat{a}_i$ will be found by minimizing some error criterion—for example, the integrated square error, C

$$C = \int \left\{ p(x \mid \omega_m) - \sum_{i=1}^{s} a_i \phi_i(x) \right\}^2 \mathrm{d}x \tag{1}$$

Differentiating C with respect to a_j yields

$$\frac{\partial C}{\partial a_j} = -2 \int \left\{ p(x \mid \omega_m) - \sum_{i=1}^{s} a_i \phi_i(x) \right\} \phi_j(x)\,\mathrm{d}x$$

$$= -2 \int p(x \mid \omega_m) \phi_j(x)\,\mathrm{d}x + 2 \sum_{i=1}^{s} a_i \int \phi_i(x)\phi_j(x)\,\mathrm{d}x$$

$$= -2 \int p(x \mid \omega_m) \phi_j(x)\,\mathrm{d}x + 2a_j$$

the last step by virtue of the orthonormality of the ϕ_i.

Equating this to zero gives

$$a_j = \int p(x \mid \omega_m) \phi_j(x)\,\mathrm{d}x$$

Although we cannot evaluate this expression directly, we can approximate it as soon as we notice that the right-hand side is merely the expected value of ϕ_j. Hence

$$\hat{a}_j = \frac{1}{n_m} \sum_{k=1}^{n_m} \phi_j(x_k) \tag{2}$$

where x_k, $k = 1, \ldots, n_m$, are the samples for class ω_m.

The final estimator of $p(x \mid \omega_m)$ is then

$$\hat{p}(x \mid \omega_m) = \sum_{i=1}^{s} \left\{ \frac{1}{n_m} \sum_{k=1}^{n_m} \phi_i(x_k) \right\} \phi_i(x)$$

One of the primary advantages of this method is that once the coefficients $\hat{a}_i$ have been determined, we can discard the sample points. Moreover, it will be apparent from the form of $\hat{a}_i$ that we could either compute $\hat{a}_i$ in one go from all the x_k, or we could compute it sequentially, in the same way that the histogram cell centres were computed in Section 2.2

$$\hat{a}_i(k+1) = \frac{1}{k+1} \{k\hat{a}_i(k) + \phi_i(x_{k+1})\}$$

This means that a very large number of sample points could be used without requiring excessive computer store or leading to long classification times. However, although we can discard the sample points, this will only lead to reduced store if s, the number of coefficients is less than n_m, the number of points. In the case when the sample points occupy a small local region of the space it is likely that a large number of terms will be needed to obtain a satisfactory approximation. One possible way to ease this difficulty is to weight the squares in (1) by $p(x)$ so that regions of higher probability density have more importance in the criterion

$$C = \int \left\{ p(x \mid \omega_m) - \sum_{i=1}^{s} a_i \phi_i(x) \right\}^2 p(x)\, \mathrm{d}x$$

This is called the integrated mean square error criterion. Unfortunately it does not have a simple elegant solution such as (2). We can even go one step further by observing that the most important region for classification is that near the decision surface and multiplying by an additional weighting factor which reaches its maximum in this region.

So far the discussion has been in terms of univariate pdfs but it can be extended quite readily to the multivariate case. A common way to define multivariate orthonormal basis functions is as products of univariate basis functions, for example

$$\Phi_{ij}(x, y) = \phi_i(x)\phi_j(y)$$

Unfortunately, however, as the number of dimensions increases so the requisite number of terms in the series increases exponentially (another instance of the curse of dimensionality). This severely limits the applicability of these methods. Readers interested in pursuing this method of density estimation will find material in Schwartz (1967), Tarter, Holcomb, and Kronmal (1967), Kronmal and Tarter (1968), and Watson (1969).

2.6 GENERAL COMMENTS

The four methods of non-parametric pdf estimation which form the basis for Sections 2.2–2.5 are the methods which have been used most often. However, other methods do exist. For example, Whittle (1958) has developed a technique in which the smoothing is determined by assuming that adjacent ordinates have a prior distribution such that they are highly correlated. Wahba (1971) uses local polynomial interpolation of the sample distribution function, differentiating the result to give an estimate of the pdf.

That all of the methods are related in some way is obvious from the fact that they all have the same aim. Nevertheless, it is of value to express these relationships explicitly. In Section 2.4 we saw that the estimate

$$\hat{p}(x \mid \omega_m) = \frac{k}{n_m V}$$

could be used by fixing V and letting k vary to give the kernel estimate, or could be used by fixing k and letting V vary to give the k-NN estimate. In Section 2.5 we expressed the series estimate in the form

$$\hat{p}(x \mid \omega_m) = \sum_{i=1}^{s} \left\{ \frac{1}{n_m} \sum_{k=1}^{n_m} \phi_i(x_k) \right\} \phi_i(x) \tag{1}$$

which, by rearranging, gives

$$\hat{p}(x \mid \omega_m) = \frac{1}{n_m} \sum_{k=1}^{n_m} \left\{ \sum_{i=1}^{s} \phi_i(x_k) \phi_i(x) \right\}$$

If we define

$$K_0\left(\frac{x - x_k}{h}\right) = h \sum_{i=1}^{s} \phi_i(x_k) \phi_i(x)$$

we see that $\hat{p}(x \mid \omega_m)$ is (2) of Section 2.3—a kernel estimate. Finally, a histogram partitions the region of support of the pdf into sets $\Gamma_1, \ldots, \Gamma_s$ and defines $\hat{p}(x \mid \omega_m)$ to be constant over Γ_i with the constant equal to

$$(\text{number of } x_k \text{ in } \Gamma_i)/n_m$$

The similarities between this and (1) of Section 2.3 are obvious. Moreover, if we let g_i be the indicator function for Γ_i (i.e. $g_i(x) = 1$ for $x \in \Gamma_i$ and 0 otherwise), then the histogram estimator is

$$\hat{p}(x \mid \omega_m) = \frac{1}{n_m} \sum_{i=1}^{s} \sum_{k=1}^{n_m} g_i(x_k) g_i(x)$$

which should be compared with (1) above.

In any particular application the practical differences rather than the theoretical differences are the determining factors in choice of method, and which practical differences are important depends, unfortunately, on the problem itself. As is so often the case, there is no 'best' method. For a high dimensional problem series methods would usually be inappropriate. If speed of classification is an important criterion then histogram methods should be considered as well as nearest-neighbour methods with reduced stored sample sets and using a branch and bound search algorithm. Careful thought should be given to whether or not peculiarities of a particular method (such as histogram discontinuities or k-NN estimators' infinite integrals) are, in fact, disadvantages. Comparisons between various methods have been made. For example, Goldstein (1975) compared kernel and k-NN methods (concluding that 'both non-parametric procedures did surprisingly well when compared with their parametric counterpart') and Anderson (1969) used analytic and Monte Carlo methods to compare the kernel and series methods. In practice, however, the most widely used method seems to be the kernel method with normal product kernels with the same h for each dimension.

2.7 FURTHER READING

Although there is undoubtedly sufficient material to fill a book on non-parametric pdf estimation, there appears as yet to be no such work of an expository nature. The nearest to it is the recent book by Wertz (1978) which presents a summary of theoretical work in the area. Although proofs are omitted it would provide an excellent source for anyone wishing to study the field. Reflecting the nature of the subject matter, a large part of the book is devoted to kernel estimators.

Fryer (1977) provides a less mathematical review, and two earlier reviews worth mentioning are those of Wegman (1972) and Cover (1972). The first presents a concise account of the mathematical properties of various estimators while the second is a less mathematical comparison.

EXERCISES

2.1 If the cells are replaced by normal distributions the Sebestyen and Edie adaptive histogram method outlined in Section 2.2 can be viewed as an approach to estimating the parameters of a normal mixture. Investigate the performance of the estimation algorithm when used in this way and compare it with the algorithms of Chapter 3.

2.2 Through simulation experiments investigate the sensitivity of the kernel classifier to choice of smoothing parameter h.

2.3 Instead of simply taking the distance to the kth nearest neighbour as the radius of the hypersphere in the k-NN method, one might take the average of the distances to the k nearest neighbours. Using analytic methods investigate the effect of such a modification on the classifier's performance.

2.4 Write a computer program for the branch and bound algorithm for finding the kth nearest neighbour in the design set from an arbitrary point $\mathbf{x}$.

2.5 In an orthogonal series estimator we can decide which terms to include by either (a) arbitrarily truncating the series at some point or (b) using a measure of goodness of fit between the model and the data and selecting a model (series) which fits well. Discuss and compare the situations in which these two approaches might be applied.

2.6 What are the advantages and disadvantages of non-parametric methods compared to parametric methods for estimating pdfs prior to classifying new observations vectors?

CHAPTER 3

Parameterized Distributions

3.1 INTRODUCTION

In Chapter 1 we considered general ways in which classifications could be made. These classifications were based on knowledge of either the class-conditional probability density functions $p(\mathbf{x}|\omega_i)$ or the class-membership probabilities $P(\omega_i|\mathbf{x})$. Since this knowledge is usually unavailable it is necessary to estimate these functions. In Chapter 2 we discussed non-parametric methods of estimating the class-conditional pdfs, i.e. methods which did not require any assumptions about the distributional forms of these pdfs. In this chapter we consider those situations where such assumptions are justifiable. In such situations advantages can result from making use of the known forms of the underlying distributions. For example, quicker classifications of new points can be made and no large database of design set points need be retained. Of course, an incorrect assumption will carry an associated cost in terms of increased error rate, but even this penalty might be acceptable if other advantages outweigh it. One could, in any case, make use of the reject option outlined in Chapter 8—using a relatively crude but quick and adequate parametric classifier over most of the space and rejecting for closer examination those points falling in some critical region where the classification was uncertain.

Although the methods of this chapter can be applied to any parameterized family of distributions we shall illustrate using two special cases: normal distributions and normal mixture distributions. There exist sound practical and theoretical reasons for proposing these forms for the role we have in mind. First let us consider class-conditional pdfs which are single normal distributions.

No student of statistics can be unaware of the central role played by the normal distribution. This is chiefly a consequence of the mathematical properties of the normal distribution. Amongst these properties are the following.

(1) The normal form is a good model of many naturally occurring phenomena. A possible reason for this may be sought in the Central Limit Theorem

which states that (when certain conditions are satisfied) a sum y of m random variables x_i $(i = 1, \ldots, m)$ tends to a normal distribution as m tends to infinity. Frequently, measurements of interest can be imagined to be a sum of many constituent components so that it comes as no surprise to discover that the measurement is approximately normally distributed. It should be made quite clear, however, that the normal distribution does not occur in nature. It is merely an abstract mathematical form which provides a good approximation to many natural distributions.

(2) The multivariate normal distribution is uniquely defined by a mean vector $\boldsymbol{\mu}$ and a variance–covariance matrix $\boldsymbol{\Sigma}$. This property has useful implications. For example, the decision surface between any two normally distributed classes will be quadratic (and linear if the classes are assumed to have identical variance–covariance matrices; see Chapter 4).

(3) After any non-singular linear transformation of the axes a normal distribution is still normal—though, of course, with different parameter values.

(4) For specified values of the mean and variance the normal distribution maximizes the entropy function. This simply means that the normal distribution is the distribution which maximizes the uncertainty of the value of an observation subject to the indicated constraints on the mean and variance of the distribution of observations.

Because the normal distribution provides a good approximation to many naturally occurring phenomena, a large amount of statistical theory has been based on it. This is particularly true of multivariate statistics. The existence of this strong body of theory predisposes us to making the assumption that a population distribution is normal. The alternative is to have some theoretical reason for proposing a particular alternative form—a fairly rare situation in multivariate problems, though less so in the univariate case. But what happens when a study of the data shows that it is clearly not normal and yet we have no alternative derived from theory? We could, of course, attempt to transform the data to provide a sample distribution which more closely follows the normal. If d was not too large one might study each of the d univariate histograms and decide if a transformation was appropriate. An alternative would be to use a mixture distribution, and in particular a normal mixture distribution, to represent each class-conditional pdf. There are two main reasons for proposing such a choice. First, a theoretical justification might be provided along the same lines as given for using a single normal distribution for each class. Thus, we might consider each class to be composed of a number of distinct types, each type having a normal population distribution. Such a situation might be felt to be realistic in, say, disease classification or machine fault identification. Even if one feels that there is no theoretical reason for choosing such a model in a particular situation, one may still accept it on the grounds that a mixture distribution will provide a better approximation than a straightforward assumption of a single distribution. One is then merely using the mixture as a way to introduce additional flexibility into the estimator—and naturally there is an associated cost in terms of the larger number of parameters which need to be estimated.

The next sections outline ways to estimate parameters in general, as well as in the particular special cases of normal distributions and normal mixtures. We begin with a discussion of the desirable properties of estimators.

3.2 DESIRABLE PROPERTIES OF ESTIMATORS

Parameter estimation procedures begin by setting up a criterion function, a function of the unknown parameters and the data, and then finding those parameter values which optimize this function. If there was only one possible criterion function which could be used, then the problem would be straightforward, but unfortunately this is not the case. It is analogous to choosing personnel on the basis of performance: do we mean performance as weightlifters or sprinters? The two are unlikely to yield the same person. Thus it is with parameter estimation: different criterion functions will yield different estimators. Fortunately, however, there are several functions which have been extensively investigated and for which the properties of the resulting estimators are well understood. Let us begin by defining a few desirable properties for estimators:

Bias. The *bias* of an estimator $\hat{\theta}$ of parameter θ is

$$\theta - E(\hat{\theta})$$

If $E(\hat{\theta}) = \theta$ then $\hat{\theta}$ is said to be *unbiased*.

Consistency. An estimator $\hat{\theta}$ of a parameter θ is said to be *consistent* if $\hat{\theta}$ converges in probability to θ. That is to say, $\hat{\theta}$ is consistent if for all $\delta, \varepsilon > 0$, there exists some N such that $n > N$ implies $P(|\hat{\theta} - \theta| < \varepsilon) > 1 - \delta$, where n is the size of the sample on which the estimate $\hat{\theta}$ is based.

Efficiency. An estimator $\hat{\theta}$ of θ is said to be *efficient* if for a given sample size it has the smallest variance of all estimators of θ.

Sufficiency. In the next section we shall make use of the joint probability density function of the sample

$$\mathscr{L}(x_1, \ldots, x_n | \theta) = \prod_{i=1}^{n} p(x_i | \theta)$$

We can factorize $\mathscr{L}$ in the perfectly general way

$$\mathscr{L}(x_1, \ldots, x_n | \theta) = g(\hat{\theta} | \theta) \mathscr{L}(x_1, \ldots, x_n | \theta, \hat{\theta})$$

The statistic $\hat{\theta}$ is then said to be a *sufficient* statistic for θ if

$$\mathscr{L}(x_1, \ldots, x_n | \theta, \hat{\theta}) = \mathscr{L}(x_1, \ldots, x_n | \hat{\theta})$$

This simply means that the joint distribution of $x_1, \ldots, x_n$, given $\hat{\theta}$, is independent of θ.

If $\hat{\theta}$ is sufficient for θ, the above factorization becomes

$$\mathscr{L}(x_1, \ldots, x_n \mid \theta) = g(\hat{\theta} \mid \theta) \mathscr{L}(x_1, \ldots, x_n \mid \hat{\theta})$$

that is, a product of two factors, one of which is a function of the data only and is independent of θ. The importance of this property will be seen below.

Identifiability. A class C of pdfs is said to be *identifiable* if and only if

$$p(x \mid \boldsymbol{\theta}) = q(x \mid \boldsymbol{\phi}) \quad \text{for all } x \quad (p, q \in C)$$

implies $\{\theta_i\} = \{\phi_j\}$, where $\{\theta_i\}$ and $\{\phi_j\}$ are, respectively, the sets with the components of $\boldsymbol{\theta}$ and $\boldsymbol{\phi}$ as elements.

Thus, the class of pdfs of the form

$$p(x \mid \theta_1, \theta_2) = \frac{1}{\theta_1 \sqrt{2\pi}} \exp\left[-\frac{1}{2} \left(\frac{x - \theta_2}{\theta_1} \right)^2 \right]$$

is identifiable since if $\phi_1 \neq \theta_1$ or $\phi_2 \neq \theta_2$ then

$$p(x \mid \theta_1, \theta_2) \neq p(x \mid \phi_1, \phi_2)$$

On the other hand, the class with elements

$$p(x \mid \theta_1, \theta_2, \theta_3) = \frac{1}{\theta_1 \sqrt{2\pi}} \exp\left[-\frac{1}{2} \left(\frac{x - \theta_2 - \theta_3}{\theta_1} \right)^2 \right]$$

is not identifiable—an infinite number of different values of θ_2 and θ_3 will give rise to the same $p(x \mid \theta_1, \theta_2, \theta_3)$.

Since we are here concerned solely with estimating the overall shape of $p(x \mid \boldsymbol{\theta})$, lack of identifiability is not a problem—any of the parameter sets giving $p(x \mid \boldsymbol{\theta})$ will do. However, the existence of multiple solutions can lead to estimation problems. Below, in Chapter 7 in the context of mixture distributions, non-identifiability could pose a very serious problem in that it would prevent estimation of the class-conditional pdfs. In general, however, the classes of pdfs we will consider will be identifiable. Nevertheless, the reader should be aware that non-identifiability can arise in circumstances more serious than the trivial example above, particularly in the context of discrete distributions and mixtures. As an example, consider:

(a) a pdf $p(x \mid \boldsymbol{\theta})$ formed from two uniform distributions, $g_1(x) = U[-2, 1]$ and $g_2(x) = U[-1, 2]$, such that

$$p(x \mid \boldsymbol{\theta}) = \tfrac{1}{2} g_1(x) + \tfrac{1}{2} g_2(x)$$

(b) a pdf $q(x \mid \boldsymbol{\theta})$ formed from two uniform distributions, $g_3(x) = U[-2, 2]$ and $g_4(x) = U[-1, 1]$, such that

$$q(x \mid \boldsymbol{\theta}) = \tfrac{2}{3} g_3(x) + \tfrac{1}{3} g_4(x)$$

Then $p(x \mid \boldsymbol{\theta}) = q(x \mid \boldsymbol{\theta})$ and a sample drawn from this distribution does not allow us to estimate the parameters uniquely. That is to say, it does not permit us to say which of (a) or (b) is the true decomposition.

Identifiability is further discussed in Teicher (1960, 1961, 1963, 1967), Rennie (1972), and Chandra (1977).

3.3 ESTIMATION METHODS

As outlined in Chapter 1, our general aim is to estimate the probability that an observed $\mathbf{x}$ comes from class ω_i. We do this via Bayes's theorem

$$P(\omega_i|\mathbf{x}) = P(\omega_i)p(\mathbf{x}|\omega_i)/p(\mathbf{x})$$

Now $p(\mathbf{x})$ is the global pdf for $\mathbf{x}$ and is independent of class. Thus

$$P(\omega_i|\mathbf{x}) \propto P(\omega_i)p(\mathbf{x}|\omega_i)$$

Also, $P(\omega_i)$, the prior probability or relative magnitude of class ω_i, is assumed to be known. The problem thus remains of how to estimate $p(\mathbf{x}|\omega_i)$, which is of known form but has unknown parameter vector $\boldsymbol{\theta}_i$. We therefore modify our notation to write $p(\mathbf{x}|\boldsymbol{\theta}_i)$ as the class-conditional pdf for class ω_i. The estimation will be based on a sample $\{\mathbf{x}_1, \ldots, \mathbf{x}_n\}$ of observations from class ω_i. That is, we want to estimate the conditional distribution

$$p(\mathbf{x}|\mathbf{x}_1, \ldots, \mathbf{x}_n)$$

The traditional way of using the information in the sample $\{\mathbf{x}_1, \ldots, \mathbf{x}_n\}$, and the way which arises from the maximum likelihood (Section 3.3.1) and distance minimization (Section 3.3.2) approaches, is to use an estimate $\hat{\boldsymbol{\theta}}_i$ in $p(\mathbf{x}|\boldsymbol{\theta}_i)$. Thus

$$p(\mathbf{x}|\mathbf{x}_1, \ldots, \mathbf{x}_n) = p(\mathbf{x}|\hat{\boldsymbol{\theta}}_i) \tag{1}$$

Aitchison *et al.* (1977) term this the *estimative* approach.

The alternative way, arising from Bayes's theorem (Section 3.3.3), is to admit that we do not know the true value of $\boldsymbol{\theta}_i$ and, instead of using a single value estimate, use a distribution $r(\boldsymbol{\theta}_i|\mathbf{x}_1, \ldots, \mathbf{x}_n)$. We can then estimate $p(\mathbf{x}|\mathbf{x}_1, \ldots, \mathbf{x}_n)$ as a weighted sum of functions $p(\mathbf{x}|\boldsymbol{\theta}_i)$ for different values of $\boldsymbol{\theta}_i$ with weights given by $r(\boldsymbol{\theta}_i|\mathbf{x}_1, \ldots, \mathbf{x}_n)$. That is

$$p(\mathbf{x}|\mathbf{x}_1, \ldots, \mathbf{x}_n) = \int p(\mathbf{x}|\boldsymbol{\theta}_i)r(\boldsymbol{\theta}_i|\mathbf{x}_1, \ldots, \mathbf{x}_n)\, d\boldsymbol{\theta}_i \tag{2}$$

Aitchison *et al.* (1977) term this the *predictive* approach. We also use the adjective *Bayesian*, since this describes the method. Note that this means that in this book 'Bayesian' is used in two ways: as above, to describe the method of estimating $p(\mathbf{x}|\mathbf{x}_1, \ldots, \mathbf{x}_n)$; and as elsewhere (e.g. Chapter 1) to describe the classification

$$\frac{P(\omega_1|\mathbf{x})}{P(\omega_2|\mathbf{x})} \gtrless k \Rightarrow \mathbf{x} \in \begin{cases} \Omega_1 \\ \Omega_2 \end{cases}$$

Taking the normal case as an example we show below that (1) leads to the normal distribution with mean and variance–covariance matrix estimated from the sample while (2) leads to a multivariate student-t type distribution. Although both of these have ellipsoidal density contours with the same centre, those of the latter are spread more sparsely. One may interpret the Bayesian method as making allowance for the sampling variability in the estimate of $\boldsymbol{\theta}_i$, just as the ordinary univariate t-test makes allowance for the extra variability introduced by replacing σ by an estimate s. This interpretation may lead one to prefer the Bayesian method. Aitchison (1975) has used divergence (see Chapter 6) to compare the performance of the methods in estimating multivariate normal pdfs—with the conclusion that the predictive method is, overall, better. Moran and Murphy (1979) have investigated this comparison further.

In the common special case of normal class-conditional pdfs with equal prior probabilities and assumed equal variance–covariance matrices, it is not difficult to see that the two methods lead to the same classification results.

3.3.1 Maximum likelihood estimation

For the moment we shall be dealing with a single class so we shall drop any class identifiers and simply refer to $p(\mathbf{x}\,|\,\boldsymbol{\theta})$ as the pdf of the class in question. $\{\mathbf{x}_1, \ldots, \mathbf{x}_n\}$ is a sample from this class. Assuming that the $\mathbf{x}_i$ have been randomly and independently selected according to $p(\mathbf{x}\,|\,\boldsymbol{\theta})$, their joint density function is

$$\mathscr{L}(\mathbf{x}_1, \ldots, \mathbf{x}_n\,|\,\boldsymbol{\theta}) = \prod_{i=1}^{n} p(\mathbf{x}_i\,|\,\boldsymbol{\theta})$$

This function, when regarded as a function of the parameters $\boldsymbol{\theta}$ rather than the data $\{\mathbf{x}_1, \ldots, \mathbf{x}_n\}$, is termed the *likelihood function*. Note that $\mathscr{L}$ is not a pdf when considered in this way. The *principle of maximum likelihood* then states that we should select as our estimate that particular $\boldsymbol{\theta}$-value which gives the greatest value of $\mathscr{L}$. This is no more than the intuitively reasonable idea that we should choose the $\boldsymbol{\theta}$-value which is most likely to give rise to the observed outcome.

Although the maximum likelihood estimate is in general not unbiased, it does possess certain desirable properties provided some fairly general conditions are satisfied. For example, under certain regularity conditions the maximum likelihood estimator is consistent. Moreover, if there exists a $\boldsymbol{\theta}^*$ which is sufficient for $\boldsymbol{\theta}$ then the maximum likelihood estimator $\hat{\boldsymbol{\theta}}$ is a function of $\boldsymbol{\theta}^*$. An extensive discussion of maximum likelihood estimators may be found in Kendall and Stuart (1961, vol. 2).

Our problem is thus one of maximizing $\mathscr{L}$ by choice of $\boldsymbol{\theta}$. Although we shall first consider the classical analytic approach to this optimization problem the reader should recognize that explicit analytic solutions are often impossible and iterative approaches (usually by computer) may have to be used. We shall consider such approaches below.

The classical approach is to differentiate $\mathscr{L}$ with respect to $\boldsymbol{\theta}$, equate the resulting expressions to 0, to give the *normal* equations

$$\partial \mathscr{L}/\partial \boldsymbol{\theta} = 0$$

and solve these equations for $\hat{\boldsymbol{\theta}}$.

As an example, suppose that the class in question has a multivariate normal pdf so that $\boldsymbol{\theta} = (\boldsymbol{\mu}, \boldsymbol{\Sigma})$ and

$$p(\mathbf{x} \mid \boldsymbol{\theta}) = \frac{1}{(2\pi)^{d/2} |\boldsymbol{\Sigma}|^{1/2}} \exp[-\tfrac{1}{2}(\mathbf{x} - \boldsymbol{\mu})' \boldsymbol{\Sigma}^{-1} (\mathbf{x} - \boldsymbol{\mu})]$$

Then

$$\mathscr{L}(\mathbf{x}_1, \ldots, \mathbf{x}_n \mid \boldsymbol{\theta}) = \frac{1}{(2\pi)^{dn/2} |\boldsymbol{\Sigma}|^{n/2}} \exp\left[-\frac{1}{2} \sum_{i=1}^{n} (\mathbf{x}_i - \boldsymbol{\mu})' \boldsymbol{\Sigma}^{-1} (\mathbf{x}_i - \boldsymbol{\mu}) \right]$$

Now clearly any monotonic transformation of $\mathscr{L}$ will possess its maximum at the same $\boldsymbol{\theta}$-value as $\mathscr{L}$ does. In particular, it is often convenient to take the natural logarithm of $\mathscr{L}$. Thus

$$L = \ln \mathscr{L} = -\frac{nd}{2} \ln 2\pi - \frac{n}{2} \ln |\boldsymbol{\Sigma}| - \frac{1}{2} \sum_{i=1}^{n} (\mathbf{x}_i - \boldsymbol{\mu})' \boldsymbol{\Sigma}^{-1} (\mathbf{x}_i - \boldsymbol{\mu})$$

so that

$$\frac{\partial L}{\partial \mu_j} = \sum_{i=1}^{n} (\mathbf{x}_i - \boldsymbol{\mu})' \boldsymbol{\Sigma}^{-1}, \qquad j = 1, \ldots, d$$

Equating this to zero gives, after a little algebra

$$\hat{\boldsymbol{\mu}} = \frac{1}{n} \sum_{i=1}^{n} \mathbf{x}_i$$

Similarly, through some slightly more complicated algebra, one can derive the maximum likelihood estimator of $\boldsymbol{\Sigma}$

$$\hat{\boldsymbol{\Sigma}} = \frac{1}{n} \sum_{i=1}^{n} (\mathbf{x}_i - \boldsymbol{\mu})(\mathbf{x}_i - \boldsymbol{\mu})'$$

In this case the differential equations could be inverted explicitly but when we turn our attention to our other illustrative special case, that of the normal mixture, we find that this is no longer true. We are therefore forced to adopt an iterative method. Since this is a necessity, one might feel tempted to abandon the normal equations and work directly with the likelihood function, using an iterative technique to find its maximum. Unfortunately, this method has a drawback: for normal mixtures the likelihood function is unbounded. We can see this as follows. The normal mixture distribution is

$$p(\mathbf{x} \mid \boldsymbol{\theta}) = \sum_{j=1}^{c} w_j f_j(\mathbf{x} \mid \boldsymbol{\mu}_j, \boldsymbol{\Sigma}_j)$$

where w_j is the relative magnitude of the jth component of the mixture, $\sum_{j=1}^{c} w_j = 1$, and where

$$f_j(\mathbf{x}\,|\,\boldsymbol{\mu}_j, \boldsymbol{\Sigma}_j) = \frac{1}{(2\pi)^{d/1}|\boldsymbol{\Sigma}_j|} \exp[-\tfrac{1}{2}(\mathbf{x} - \boldsymbol{\mu}_j)\boldsymbol{\Sigma}_j^{-1}(\mathbf{x} - \boldsymbol{\mu}_j)]$$

The likelihood function is thus

$$\mathscr{L} = \prod_{i=1}^{n} p(\mathbf{x}_i\,|\,\boldsymbol{\theta}) = \prod_{i=1}^{n}\left[\sum_{j=1}^{c} w_j f_j(\mathbf{x}_i\,|\,\boldsymbol{\mu}_j, \boldsymbol{\Sigma}_j)\right]$$

$$= \left[\sum_{j=1}^{c} w_j f_j(\mathbf{x}_1\,|\,\boldsymbol{\mu}_j \boldsymbol{\Sigma}_j)\right]\left[\prod_{i=2}^{n}\sum_{j=1}^{c} w_j f_j(\mathbf{x}_i\,|\,\boldsymbol{\mu}_j, \boldsymbol{\Sigma}_j)\right]$$

Now the right-hand factor is greater than or equal to

$$\prod_{i=2}^{n}\sum_{j=2}^{c} w_j f_j(\mathbf{x}_i\,|\,\boldsymbol{\mu}_j, \boldsymbol{\Sigma}_j) = k$$

(where the first component of the mixture has been dropped). So

$$\mathscr{L} \geqslant k \sum_{j=1}^{c} w_j f_j(\mathbf{x}_1\,|\,\boldsymbol{\mu}_j, \boldsymbol{\Sigma}_j)$$

$$\geqslant k w_1 f_1(\mathbf{x}_1\,|\,\boldsymbol{\mu}_1, \boldsymbol{\Sigma}_1)$$

$$= k w_1 \frac{1}{(2\pi)^{d/2}|\boldsymbol{\Sigma}_1|^{1/2}} \exp[-\tfrac{1}{2}(\mathbf{x}_1 - \boldsymbol{\mu}_1)'\boldsymbol{\Sigma}_1^{-1}(\mathbf{x}_1 - \boldsymbol{\mu}_1)]$$

Now estimate $\boldsymbol{\mu}_1$ by $\hat{\boldsymbol{\mu}}_1 = \mathbf{x}_1$. Then

$$\mathscr{L} \geqslant k w_1/(2\pi)^{d/2}|\boldsymbol{\Sigma}_1|^{1/2}$$

If now $|\boldsymbol{\Sigma}_1| \to 0$ then $\mathscr{L} \to \infty$.

The implication of this is that if we tried to maximize the likelihood function directly by hill climbing we could converge to a singularity which is useless for our purpose. Recognizing this danger, some authors have constrained the covariance matrices of the components to be proportional to each other. Others have suggested finding the largest local maximum, i.e. we could try to find the solution of the normal equations $\partial\mathscr{L}/\partial\boldsymbol{\theta} = 0$ which gives the largest value of $\mathscr{L}$. We are thus back at the classical solution, although the impossibility of inverting the normal equations means that we must still use an iterative technique, only now to solve $\partial\mathscr{L}/\partial\boldsymbol{\theta} = 0$ instead of to maximize $\mathscr{L}$. It should be noted that convergence of iterative methods is not usually guaranteed under all circumstances so that even this method does not guarantee that the solution vector will not approach a singularity of $\mathscr{L}$.

One iterative technique which is commonly used for normal mixtures cycles repeatedly through two stages, the first of which estimates the membership probability for the jth normal component at $\mathbf{x}_i$ by

$$\hat{F}(j \mid \mathbf{x}_i) = \frac{\hat{w}_j f_j(\mathbf{x}_i \mid \hat{\boldsymbol{\mu}}_j, \hat{\boldsymbol{\Sigma}}_j)}{\hat{p}(\mathbf{x}_i)}$$

(where $\hat{p}(\mathbf{x}_i) = \Sigma_j \hat{w}_j f_j(\mathbf{x}_i \mid \hat{\boldsymbol{\mu}}_j, \hat{\boldsymbol{\Sigma}}_j)$) and the second of which uses these membership probability estimates to update the estimates of w_j, $\boldsymbol{\mu}_j$, and $\boldsymbol{\Sigma}_j$. Thus

$$\hat{w}_j = \frac{1}{n} \sum_{i=1}^{n} \hat{F}(j \mid \mathbf{x}_i)$$

$$\hat{\boldsymbol{\mu}}_j = \frac{1}{n\hat{w}_j} \sum_{i=1}^{n} \hat{F}(j \mid \mathbf{x}_i)\mathbf{x}_i$$

$$\hat{\boldsymbol{\Sigma}}_j = \frac{1}{n\hat{w}_j} \sum_{i=1}^{n} \hat{F}(j \mid \mathbf{x}_i)(\mathbf{x}_i - \hat{\boldsymbol{\mu}}_j)(\mathbf{x}_i - \hat{\boldsymbol{\mu}}_j)'$$

These updated parameter estimates are then substituted back to give revised estimates of $\hat{F}(j \mid \mathbf{x}_i)$. This kind of algorithm has been suggested by Behboodian (1970), Wolfe (1969), and Hasselblad (1966). Everitt and Hand (1981) illustrate its application to mixtures of non-normal components.

An example of this approach is given in Figures 3.1, 3.2, and 3.3. Here the data (shown in Figure 3.1) consist of measurements of the antenna lengths and cauda widths of 52 aphids. It was thought that a mixture of two normal distributions would adequately describe the distribution. Figure 3.2 shows the result when equal covariance matrices are assumed for the two classes. Figure 3.3 is the same, but when unequal variance-covariance matrices are permitted.

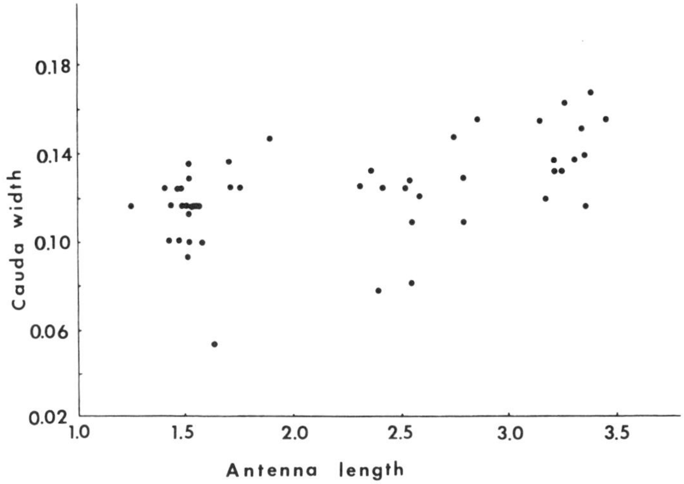

Figure 3.1 Antenna length by cauda width (mm) for 52 aphids (some points occur on top of each other)

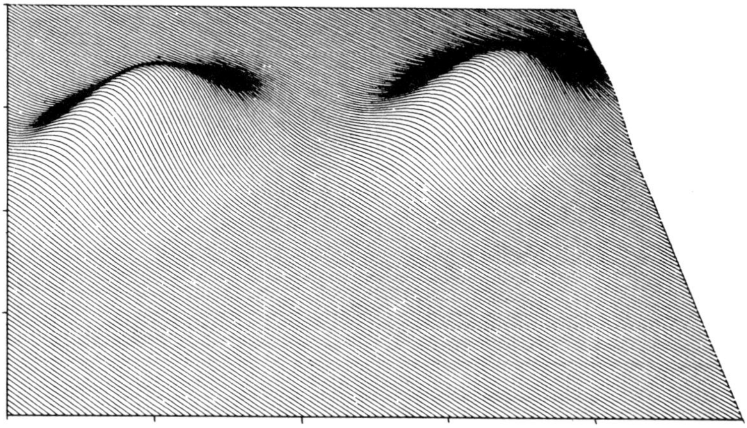

Figure 3.2 Fitting the data of Figure 3.1 by a two-component normal mixture (with equal variance–covariance matrices) using Wolfe's (1969) iterative maximum likelihood solution

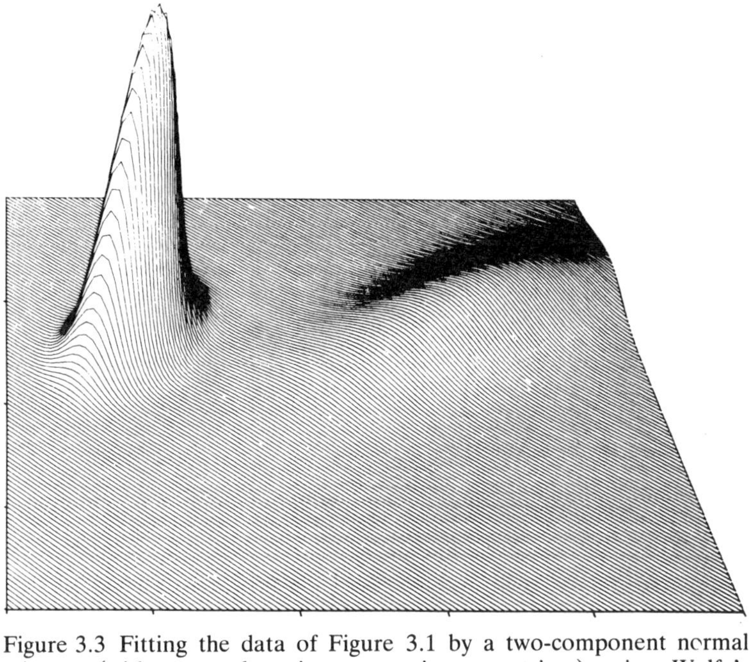

Figure 3.3 Fitting the data of Figure 3.1 by a two-component normal mixture (with unequal variance–covariance matrices) using Wolfe's (1969) iterative maximum likelihood solution

3.3.2 Distance minimization

If, in a univariate problem, we knew the true density function $p(y)$ we could plot both this and, for a given parameter vector, the estimated function $p^*(y\,|\,\boldsymbol{\theta})$. We could then estimate the parameters of the latter by choosing those values which minimized the difference between the two distributions. This ideal approach has two drawbacks. The first is that, of course, we do not know $p(y)$: we have only a sample from $p(y)$. The second is that the method cannot be applied in multivariate situations, with which we are chiefly concerned. In this section we discuss estimators which are based on this fundamental approach. We show how to use the observed sample to replace $p(y)$ and how to generalize the ideas to many variables.

All of the estimators in this section have two steps in common: they begin by defining a 'distance' measure between the sample and $p^*(y\,|\,\boldsymbol{\theta})$ and they then choose the parameters $\boldsymbol{\theta}$ to minimize the distance. Many distance measures have been proposed for use in this context, and they differ in two ways. First, they differ in the features they use to describe the sample and the theoretical curve. Secondly, they differ in the way in which these features are combined to form a single distance measure which can be minimized.

We begin with a method which is not very different from the fundamental approach by using a non-parametric estimator, say the kernel method of the previous chapter, to provide an initial estimate of the class-conditional pdf (recall that we are here restricting the discussion to a single class). This estimate $\hat{p}(\mathbf{y}\,|\,\mathbf{x}_1, \ldots, \mathbf{x}_n)$ replaces $p(\mathbf{y})$ in the above. We could then use the integrated squared error between this non-parametric estimate and the parametric estimate, $p^*(\mathbf{y}\,|\,\boldsymbol{\theta})$ as a distance function, choosing that parameter vector which minimized the integrated squared error. Thus, the distance function is

$$D_1(\boldsymbol{\theta}) = \int [\hat{p}(\mathbf{y}\,|\,\mathbf{x}_1, \ldots, \mathbf{x}_n) - p^*(\mathbf{y}\,|\,\boldsymbol{\theta})]^2 \, d\mathbf{y} \tag{1}$$

Finding the optimum value of $\boldsymbol{\theta}$ using this function will be difficult, especially in the multivariate case where multiple integrals will be needed. We therefore replace the integrated squared error by a sum of squares

$$D_2(\boldsymbol{\theta}) = \sum_{i=1}^{m} [\hat{p}(\mathbf{y}_i|\mathbf{x}_1, \ldots, \mathbf{x}_n) - p^*(\mathbf{y}_i|\boldsymbol{\theta})]^2 \tag{2}$$

Here $\{\mathbf{y}_1, \ldots, \mathbf{y}_m\}$ is a set of test points. Clearly, the larger m is the closer the solution obtained using D_2 is to that which would be obtained if D_1 was used. Equally clearly, this imposes a limitation of the method in multivariate problems since, for large d, m would need to be very large so that the test points adequately covered the space of $\mathbf{y}$.

In this approach the features are the values of $\hat{p}(\mathbf{y}_i|\mathbf{x}_1, \ldots, \mathbf{x}_n)$ and $p^*(\mathbf{y}_i|\boldsymbol{\theta})$ $(i = 1, \ldots, m)$. One need not think of the non-parametric estimate at all, but can simply regard the $\hat{p}(\mathbf{y}_i|\mathbf{x}_1, \ldots, \mathbf{x}_n)$ as certain values estimated from the sample, which are compared with matching values, $p^*(\mathbf{y}_i|\boldsymbol{\theta})$, derived as functions of $\boldsymbol{\theta}$.

One could equally apply this approach to cumulative distribution functions, characteristic functions, or moment-generating functions. Examples of these three realizations are given in, respectively, Choi and Bulgren (1968), Binder (1978), and Quandt and Ramsey (1978), who use the methods to estimate the parameters of mixture distributions.

In (2) above we modified (1) by taking the values of $\hat{p}(\mathbf{y}|\mathbf{x}_1, \ldots, \mathbf{x}_n)$ $p^*(\mathbf{y}|\boldsymbol{\theta})$ at m test points $\mathbf{y}_i$. This ignores the values of these functions at all other $\mathbf{y}$-values. We could try to take all $\mathbf{y}$-values into account by dividing the region of $\mathbf{y}$-space we are interested in into m non-overlapping cells which cover the entire region and comparing the areas under $\hat{p}(\mathbf{y}|\mathbf{x}_1, \ldots, \mathbf{x}_n)$ and $p^*(\mathbf{y}|\boldsymbol{\theta})$ for each of the cells. The area under $\hat{p}(\mathbf{y}|\mathbf{x}_1, \ldots, \mathbf{x}_n)$ can be estimated by $N_i(\mathbf{x}_1, \ldots, \mathbf{x}_n)$, the proportion of the n observations falling in cell i. The distance function is then

$$D_3(\boldsymbol{\theta}) = \sum_{i=1}^{m} \left[N_i(\mathbf{x}_1, \ldots, \mathbf{x}_n) - \int_i p^*(\mathbf{y}|\boldsymbol{\theta})\, d\mathbf{y} \right]^2$$

where $\int_i$ signifies integration over cell i. Indeed, if this approach is adopted one is almost using the minimum χ^2 criterion

$$D_4(\boldsymbol{\theta}) = \sum_{i=1}^{m} \left[N_i(\mathbf{x}_1, \ldots, \mathbf{x}_n) - \int_i p^*(\mathbf{y}|\boldsymbol{\theta})\, d\mathbf{y} \right]^2 \Big/ \int_i p^*(\mathbf{y}|\boldsymbol{\theta})\, d\mathbf{y}$$

(See, for example, Kendall and Stuart, 1961, vol. 2.)

All of the comments in Chapter 2 about histograms in high dimensional spaces apply here.

Note also that multiple integrals over $p^*(\mathbf{y}|\boldsymbol{\theta})$ are required. Again the features can be thought of simply as functions of the sample and $p^*(\mathbf{y}|\boldsymbol{\theta})$ rather than as derived from an intermediary step of pdf estimation.

The minimum χ^2 method introduced in D_4 weights each of the feature differences by a term $[\int_i p^*(\mathbf{y}|\boldsymbol{\theta})\, d\mathbf{y}]^{-1}$. One could equally well use other weights. In particular, one could use a weight of $p(\mathbf{y})$ in (1) to give the mean square error

$$\int p(\mathbf{y})[\hat{p}(\mathbf{y}|\mathbf{x}_1, \ldots, \mathbf{x}_n) - p^*(\mathbf{y}|\boldsymbol{\theta})]^2\, d\mathbf{y}$$

approximated by

$$\sum_{i=1}^{m} \hat{p}(\mathbf{y}_i|\mathbf{x}_1, \ldots, \mathbf{x}_n)[\hat{p}(\mathbf{y}_i|\mathbf{x}_1, \ldots, \mathbf{x}_n) - p^*(\mathbf{y}_i|\boldsymbol{\theta})]^2$$

Similarly one could use a more general system of weighting, replacing the $\hat{p}(\mathbf{y}_i|\mathbf{x}_1, \ldots, \mathbf{x}_n)$ weights by costs $c(\mathbf{y}_i)$—or, indeed, use both.

A second way to generalize (1) is to take a power other than 2. If we let r be an arbitrary exponent and take the rth root of the sum to keep the dimensions the same, we have, in general

$$D_5(\boldsymbol{\theta}) = [\int [\hat{p}(\mathbf{y}|\mathbf{x}_1, \ldots, \mathbf{x}_n) - p^*(\mathbf{y}|\boldsymbol{\theta})]^r\, d\mathbf{y}]^{1/r}$$

With $r = 2$ this is basically D_1, and with $r = \infty$ it seeks to minimize the maximum deviation between $\hat{p}(\mathbf{y} \mid \mathbf{x}_1, \ldots, \mathbf{x}_n)$ and $p^*(\mathbf{y} \mid \boldsymbol{\theta})$.

Yet another way to generalize is to replace $\hat{p}(\mathbf{y} \mid \mathbf{x}_1, \ldots, \mathbf{x}_n)$ and $p^*(\mathbf{y} \mid \boldsymbol{\theta})$ by functions of these. Thus, Young and Coraluppi (1970) use a particularly interesting criterion of this type based on the information function, namely

$$\int p(\mathbf{y})[\ln p(\mathbf{y}) - \ln p^*(\mathbf{y} \mid \boldsymbol{\theta})]^2 \, d\mathbf{y}$$

This function is discussed further in Section 3.4.

We have assumed implicitly in the above that, with l parameters to be estimated and m pairs of features (one of each pair being from the sample and one from $p^*(\mathbf{y} \mid \boldsymbol{\theta})$), $m > l$. But what if $m = l$? Each feature pair can be used to set up an equation relating $\boldsymbol{\theta}$, through $p^*(\mathbf{y} \mid \boldsymbol{\theta})$, to some function of the data. We thus have a system of m equations in m unknowns, with the apparent possibility of solving the system for the m parameters. (In effect the minimum of the distance function is zero, and the $\boldsymbol{\theta}$-values which give this minimum are found.)

This sort of approach is well established, having been widely used before computers (which permit ready minimization of complicated functions) became available. Again there is a very wide choice of precisely what feature set to use. Kabir (1968) estimates mixture distribution parameters using proportions falling in cells as features. Bartholomew (1959) does the same but using proportions in half open intervals. The most widely applied method of this type is, however, probably the method of moments. This uses the first few sample and theoretical moments as features.

Of course, the feasibility of inverting the system of equations by hand does not imply that this is simple—as the algebra of Pearson (1894) demonstrates (see also Everitt and Hand, 1981) when the method is applied to a mixture of two normal components. The algebraic complexity of even this simple case suggests that the method will have severe practical limitations in our application. There is, however, an additional problem which arises when a large number of parameters have to be estimated. This is simply that the sampling variance for higher moments becomes very large so that inaccurate estimates result. Day (1969) has studied two-component multivariate normal mixtures and found that in two or more dimensions the method does not perform well.

One other point should be made about this estimation approach. Unless a simple form is adopted for the class-conditional pdf, the system of m equations to be inverted will not be linear and may lead to more than one possible solution. In such an event it is usual to choose a solution on the basis of the next higher moment. Note that in doing this one is in effect defining a very complicated distance function based on the first $(m + 1)$ moments as features. This function combines the $(m + 1)$ feature differences in such a way that the first m are weighted extremely heavily relative to the $(m + 1)$st. Thus, the parameter estimates make the first m feature differences zero and minimize the $(m + 1)$st.

Figure 3.4 illustrates a distance minimization algorithm applied to the aphid

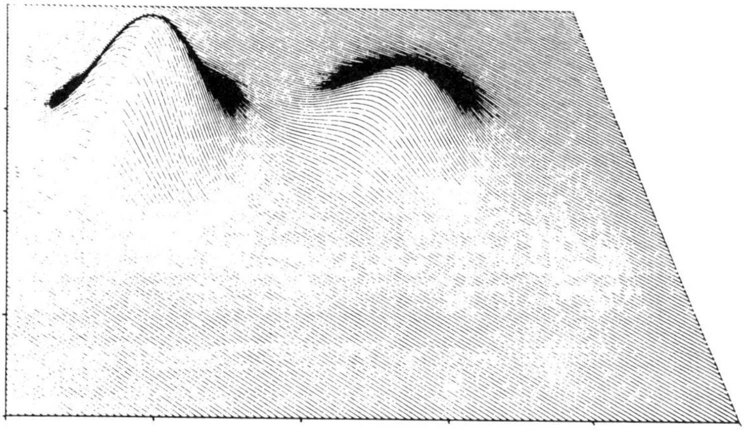

Figure 3.4 Fitting the data of Figure 3.1 by a two-component normal mixture (with equal variance–covariance matrices) by minimizing a sum of squared differences between a non-parametric (kernel) estimate and the mixture distribution

data of Figure 3.1. The function minimized here was the sum of squares of the differences between a non-parametric kernel estimate and the parameterized mixture distribution, evaluated on a rectangular array of 100 $\mathbf{y}_i$ points.

3.3.3 Bayes methods

In the introductory paragraphs to this section we stated that Bayes (predictive) methods do not base the estimate of density $p(\mathbf{x} | \mathbf{x}_1, \ldots, \mathbf{x}_n)$ on a simple estimate of $\boldsymbol{\theta}$, but rather form a weighted combination of densities $p(\mathbf{x} | \boldsymbol{\theta})$ with weights given by a function $r(\boldsymbol{\theta} | \mathbf{x}_1, \ldots, \mathbf{x}_n)$. (Note that for convenience we have again temporarily dropped any class identifiers.) We then have

$$p(\mathbf{x} | \mathbf{x}_1, \ldots, \mathbf{x}_n) = \int p(\mathbf{x} | \boldsymbol{\theta}) r(\boldsymbol{\theta} | \mathbf{x}_1, \ldots, \mathbf{x}_n) \, d\boldsymbol{\theta} \tag{1}$$

and $r(\boldsymbol{\theta} | \mathbf{x}_1, \ldots, \mathbf{x}_n)$ may be regarded as a density function over possible values of $\boldsymbol{\theta}$. For this approach our problem has become that of estimating the function $r(\boldsymbol{\theta} | \mathbf{x}_1, \ldots, \mathbf{x}_n)$. This is done by starting from an initial *prior* density $r(\boldsymbol{\theta})$, which may be regarded as a measure of confidence of our belief that $\boldsymbol{\theta}$ has each value. $r(\boldsymbol{\theta})$ is then updated to take into account the observed sample $\{\mathbf{x}_1, \ldots, \mathbf{x}_n\}$. Thus, by Bayes's theorem

$$r(\boldsymbol{\theta} | \mathbf{x}_1, \ldots, \mathbf{x}_n) = \frac{\mathscr{L}(\mathbf{x}_1, \ldots, \mathbf{x}_n | \boldsymbol{\theta}) r(\boldsymbol{\theta})}{\int \mathscr{L}(\mathbf{x}_1, \ldots, \mathbf{x}_n | \boldsymbol{\theta}) r(\boldsymbol{\theta}) \, d\boldsymbol{\theta}} \tag{2}$$

where $r(\boldsymbol{\theta} | \mathbf{x}_1, \ldots, \mathbf{x}_n)$ is called the *posterior* density since it is the density of $\boldsymbol{\theta}$ after taking the observed data into consideration.

It is obvious that in the general case evaluation of the posterior density will be a lengthy task. Fortunately, for many special cases $r(\boldsymbol{\theta}|\mathbf{x}_1, \ldots, \mathbf{x}_n)$ takes the same form as $r(\boldsymbol{\theta})$ and one can estimate the posterior density merely by replacing the parameters in the prior. In such a case the density of $\boldsymbol{\theta}$ is called a *reproducing density*. Spragins (1965) proves that a reproducing density $r(\boldsymbol{\theta})$ exists if and only if the observations $\mathbf{x}_1, \ldots, \mathbf{x}_n$ admit a sufficient statistic (Section 3.2) expressible as a vector of fixed dimension.

As we illustrate below, normal distributions possess this simplifying property but unfortunately normal mixtures do not do so, with the result that difficulties are encountered when trying to apply the method to normal mixtures. One might try to tackle this by using numerical techniques (perhaps Monte Carlo evaluation of the integrals) or one might makc simplifying assumptions.

To illustrate the above theory we begin with the very simple case of a univariate normal distribution with standard deviation σ assumed known and mean μ to be estimated. Our aim is thus to find $r(\mu|x_1, \ldots, x_n)$. As the prior distribution of μ we take a normal distribution with mean μ_0 and variance σ_0^2. Thus

$$r(\mu) = \frac{1}{\sigma_0\sqrt{2\pi}} \exp\left[-\frac{1}{2}\left(\frac{\mu-\mu_0}{\sigma_0}\right)^2\right]$$

We can rewrite (2) in the form

$$r(\boldsymbol{\theta}|\mathbf{x}_1, \ldots, \mathbf{x}_n) = k(\mathbf{x}_1, \ldots, \mathbf{x}_n) r(\boldsymbol{\theta}) \prod_{i=1}^{n} p(\mathbf{x}_i|\boldsymbol{\theta})$$

which in the current special case becomes

$$\begin{aligned} r(\mu|x_1, \ldots, x_n) &= k(x_1, \ldots, x_n) \frac{1}{\sigma_0\sqrt{2\pi}} \exp\left[-\frac{1}{2}\left(\frac{\mu-\mu_0}{\sigma_0}\right)^2\right] \\ &\quad \times \prod_{i=1}^{n} \frac{1}{\sigma\sqrt{2\pi}} \exp\left[-\frac{1}{2}\left(\frac{x_i-\mu}{\sigma}\right)^2\right] \\ &= \frac{k(x_1, \ldots, x_n)}{\sigma_0\sigma^n(2\pi)^{(n+1)/2}} \exp\left[-\frac{1}{2}\sum_{i=1}^{n}\left(\frac{x_i-\mu}{\sigma}\right)^2 - \frac{1}{2}\left(\frac{\mu-\mu_0}{\sigma_0}\right)^2\right] \end{aligned}$$

Some algebra allows us to re-write the exponent of this expression as

$$-\frac{1}{2}\left[\left(\frac{\mu-\mu_n}{\sigma_n}\right)^2 - k_n\right]$$

where

$$\mu_n = \left[\frac{\Sigma x_i}{\sigma^2} + \frac{\mu_0}{\sigma_0^2}\right] \Big/ \left[\frac{n}{\sigma^2} + \frac{1}{\sigma_0^2}\right]$$

$$\sigma_n^2 = \left[\frac{n}{\sigma^2} + \frac{1}{\sigma_0^2}\right]^{-1}$$

and

$$k_n = \left[\frac{\Sigma x_i}{\sigma^2} + \frac{\mu_0}{\sigma_0^2}\right]^2 \bigg/ \left[\frac{n}{\sigma^2} + \frac{1}{\sigma_0^2}\right] + \left[\frac{\Sigma x_i^2}{\sigma^2} + \frac{\mu_0^2}{\sigma_0^2}\right]$$

so that we can put

$$r(\mu|x_1, \ldots, x_n) = k_1 \frac{1}{\sigma_n\sqrt{2\pi}} \exp\left[-\frac{1}{2}\left(\frac{\mu - \mu_n}{\sigma_n}\right)^2\right]$$

where k_1 is a function of $\mathbf{x}_1, \ldots, \mathbf{x}_n, \sigma, \sigma_0, n$, and μ_0 but not of μ. This means that the posterior distribution is simply a normal distribution with mean μ_n and standard deviation σ_n. Note that since $r(\mu|x_1, \ldots, x_n)$ is a pdf its integral is necessarily unity so $k_1 = 1$.

The forms of μ_n and σ_n^{-2} are particularly revealing. We have

$$\mu_n = \frac{n\sigma_0^2}{n\sigma_0^2 + \sigma^2}\bar{x} + \frac{\sigma^2}{n\sigma_0^2 + \sigma^2}\mu_0$$

which is simply a linear combination of the sample mean $\bar{x}$ and the initial guess at the mean μ_0. As n increases so the contribution due to $\bar{x}$ becomes proportionately greater. Similarly

$$\frac{1}{\sigma_n^2} = \frac{n}{\sigma^2} + \frac{1}{\sigma_0^2}$$

is a weighted combination of $1/\sigma^2$ and $1/\sigma_0^2$ and as n increases the relative contribution from the prior decreases and

$$\frac{1}{\sigma_n^2} \to \frac{n}{\sigma^2}$$

That is, $\sigma_n^2 \to \sigma^2/n$, the variance of $\bar{x}$.

If we now substitute the normal density for $p(x|\mu)$ and the posterior density for μ derived above in (1) we arrive at

$$p(x|x_1, \ldots, x_n) = \frac{1}{(\sigma^2 + \sigma_n^2)^{1/2}\sqrt{2\pi}} \exp\left[-\frac{1}{2}\frac{(x - \mu_n)^2}{\sigma^2 + \sigma_n^2}\right]$$

so that the final density function of x given $x_1, \ldots, x_n$ is also normal.

The example above shows how the method is applied in practice. It may also suggest that the algebraic manipulations involved in arriving at the posterior distribution can be quite involved. In any case the example is of little use for our application since we would normally have $d > 1$ and σ^2 also unknown. We therefore give, without proof, the result for a multivariate normal distribution with both $\boldsymbol{\mu}$ and $\boldsymbol{\Sigma}$ unknown. Readers interested in the derivation of this result may refer to Keehn (1965) and also to De Groot (1970). The form we give follows Keehn and others in expressing the distributions in terms of $\mathbf{P}$, the inverse of $\boldsymbol{\Sigma}$, rather than in terms of $\boldsymbol{\Sigma}$ itself (the advantages of this are hinted at by the role σ^{-2} plays in the above one dimensional derivation). As the prior density we take the distribution of the covariance matrix to be an inverted Wishart law with parameters ν_0 and $\boldsymbol{\phi}_0$ and the distribution of the mean to be normal with mean $\boldsymbol{\mu}_0$ and covariance matrix $\mathbf{P}^{-1}/w_0$. (w_0 reflects the initial confidence in $\boldsymbol{\mu}_0$ as the value for the mean.) The prior density is thus

$$r(\boldsymbol{\mu}, \mathbf{P}) = (2\pi)^{-d/2}|w_0\mathbf{P}|^{1/2} \exp[\tfrac{1}{2}w_0(\boldsymbol{\mu} - \boldsymbol{\mu}_0)'\mathbf{P}(\boldsymbol{\mu} - \boldsymbol{\mu}_0)] \times c(d, \nu_0)\left|\frac{\nu_0}{2}\boldsymbol{\phi}_0\right|^{(\nu_0-1)/2} |\mathbf{P}|^{(\nu_0-d-2)/2} \exp[-\tfrac{1}{2} \operatorname{tr} \nu_0\boldsymbol{\phi}_0\mathbf{P}]$$

where

$$c(d, \nu_0) = \left[\pi^{d(d-1)/4} \prod_{i=1}^{d} \Gamma\left(\frac{\nu_0 - i}{2}\right)\right]^{-1}.$$

Multiplying by the likelihood function for n independent samples drawn from the d-dimensional normal distribution with parameters $\boldsymbol{\mu}$ and $\mathbf{P}$ leads to the posterior density

$$r(\boldsymbol{\mu}, \mathbf{P}|\mathbf{x}_1, \ldots, \mathbf{x}_n) = (2\pi)^{-d/2}|w_n\mathbf{P}|^{1/2} \exp[-\tfrac{1}{2}w_n(\boldsymbol{\mu} - \boldsymbol{\mu}_n)'\mathbf{P}(\boldsymbol{\mu} - \boldsymbol{\mu}_n)] \times c(d, \nu_n)\left|\frac{\nu_n}{2}\boldsymbol{\phi}_n\right|^{(\nu_n-1)/2} |\mathbf{P}|^{(\nu_n-d-2)/2} \times \exp[-\tfrac{1}{2} \operatorname{tr} \nu_n\boldsymbol{\phi}_n\mathbf{P}]$$

where

$$w_n = w_0 + n$$
$$\nu_n = \nu_0 + n$$
$$\boldsymbol{\mu}_n = (w_0\boldsymbol{\mu}_0 + n\bar{\mathbf{x}})/(w_0 + n)$$

and

$$\boldsymbol{\phi}_n = \frac{\nu_0\boldsymbol{\phi}_0 + w_0\boldsymbol{\mu}_0\boldsymbol{\mu}_0' + (n - 1.)\mathbf{S} + n\overline{\mathbf{x}\mathbf{x}}' - w_n\boldsymbol{\mu}_n\boldsymbol{\mu}_n'}{\nu_0 + n}$$

with

$$\mathbf{S} = \frac{1}{n - 1} \sum_{i=1}^{n} (\mathbf{x}_i - \bar{\mathbf{x}})(\mathbf{x}_i - \bar{\mathbf{x}})'$$

the sample variance–covariance matrix, and

$$\bar{\mathbf{x}} = \frac{1}{n} \sum_{i=1}^{n} \mathbf{x}_i$$

the sample mean. Keehn (1965) then goes on to derive

$$p(\mathbf{x}|\mathbf{x}_1, \ldots, \mathbf{x}_n) = \frac{(2\pi)^{-d/2}\left|\dfrac{\nu_n}{2}\boldsymbol{\phi}_n\right|^{-1/2} \Gamma\left(\dfrac{\nu_n}{2}\right)\Big/\Gamma\left(\dfrac{\nu_n - d}{2}\right)}{\left\{1 + \dfrac{w_n + 1}{w_n} \cdot \dfrac{1}{\nu_n} \operatorname{tr}[(\mathbf{x} - \boldsymbol{\mu}_n)(\mathbf{x} - \boldsymbol{\mu}_n)'\boldsymbol{\phi}_n^{-1}]\right\}^{\nu_n/2}}$$

which is the d-dimensional generalization of Student's t-distribution referred to above.

Further work on the Bayesian approach to classification with normal class conditional pdfs may be found in Geisser (1966) and Chien and Fu (1967). More general work on Bayesian estimation is given in Lindley (1965) and Jaynes (1968), the latter who tackles the thorny problem of choice of prior. Hora (1978) has studied the determination of optimal sample sizes for Bayesian discriminant analysis.

3.4 RELATIONSHIPS BETWEEN THE METHODS

Superficially the three parameter estimation methods discussed in Section 3.3 are very different. The first estimates $\boldsymbol{\theta}$ by maximizing the likelihood function while the second estimates $\boldsymbol{\theta}$ by minimizing an error criterion. The third method does not even estimate $\boldsymbol{\theta}$, but rather yields a density function over $\hat{\boldsymbol{\theta}}$. Nevertheless the three approaches have the same aims so one would expect certain relationships to exist.

First let us consider maximum likelihood and distance minimization. As outlined in Sections 3.3.2 and 3.5, Young and Coraluppi (1970) use as their distance criterion the information function

$$D(\boldsymbol{\theta}) = \int p(x)[\ln p(x) - \ln p^*(x \mid \boldsymbol{\theta})] \, dx$$

measuring the error resulting from estimating the true pdf $p(x)$ by $p^*(x \mid \boldsymbol{\theta})$. We can rewrite this as

$$D(\boldsymbol{\theta}) = \int p(x) \ln p(x) \, dx - \int p(x) \ln p^*(x \mid \boldsymbol{\theta}) \, dx$$

the first term of which is a constant so that minimizing $D(\boldsymbol{\theta})$ is equivalent to maximizing

$$\int p(x) \ln p^*(x \mid \boldsymbol{\theta}) \, dx$$

This is simply the expected value of $\ln p^*(x \mid \boldsymbol{\theta})$. In Section 3.3.2 we estimated the true pdf by $\hat{p}(x)$. We could do that here and evaluate the integral, or we could simply estimate the expected value directly from the sample as

$$\frac{1}{n} \sum_{i=1}^{n} \ln p^*(x_i \mid \boldsymbol{\theta}) = \frac{1}{n} \ln \prod_{i=1}^{n} p^*(x_i \mid \boldsymbol{\theta})$$

If we assume that the true pdf is indeed from the family $p^*(x \mid \boldsymbol{\theta})$ then the final expression is simply $1/n$ times the log-likelihood.

This has demonstrated a relationship between the maximum likelihood method and the distance minimization method for a certain distance function. Now let us consider the maximum likelihood and Bayes methods.

In the Bayes method the *a posteriori* distribution of $\boldsymbol{\theta}$ is

$$p(\boldsymbol{\theta} \mid x_1, \ldots, x_n) \propto \mathscr{L}(x_1, \ldots, x_n \mid \boldsymbol{\theta}) p(\boldsymbol{\theta})$$

If there is little *a priori* information then $p(\boldsymbol{\theta})$ is not zero and is almost flat in the region of interest. In particular, $p(\boldsymbol{\theta})$ is almost flat in the vicinity of $\hat{\boldsymbol{\theta}}$, the maximum likelihood solution. Furthermore, by definition $\mathscr{L}(x_1, \ldots, x_n | \boldsymbol{\theta})$, the likelihood function, peaks at $\hat{\boldsymbol{\theta}}$, the maximum likelihood solution. Thus, $p(\boldsymbol{\theta} | x_1, \ldots, x_n)$ will also peak near $\hat{\boldsymbol{\theta}}$.

The posterior distribution of x is then found from

$$p(x | x_1, \ldots, x_n) = \int p(x | \boldsymbol{\theta}) p(\boldsymbol{\theta} | x_1, \ldots, x_n) \, d\boldsymbol{\theta}$$

and this will be approximately equal to $p(x | \hat{\boldsymbol{\theta}})$ if the peak of $p(\boldsymbol{\theta} | x_1, \ldots, x_n)$ at $\hat{\boldsymbol{\theta}}$ is sharp. Thus, the Bayes solution will be approximately equal to the maximum likelihood solution.

For this result we had to assume that $p(\boldsymbol{\theta})$ was flat (and not zero) and that the peak of $\mathscr{L}$ at $\hat{\boldsymbol{\theta}}$ was sharp. To the extent that these assumptions do not hold the two solutions will differ.

3.5 SEQUENTIAL METHODS OF PARAMETER ESTIMATION

Sequential or adaptive methods are important in pattern recognition for several reasons. First there is the historical perspective of attempts to model human brain function. Objects or observation vectors are presented one at a time and classification rules are gradually built up—the system 'learns' to classify new points. Secondly, very large sets of data can be made full use of in this way. As each point is presented the classifier is updated to take it into account, after which the point can be discarded. This avoids the need to retain very large databases. Thirdly, by making use of classifiers which can adjust themselves should the nature of the incoming points change, dynamic or time-varying class structures can be handled.

Sequential methods are important in the theory of linear decision surfaces (Chapter 4), in cluster analysis (Chapter 7), and, in a rather different way, in variable selection (Chapter 6). In this section we concern ourselves solely with sequential estimation of the parameters of class-conditional pdfs.

We shall begin by considering the maximum likelihood solution for the single normal distribution given in Section 3.3.1. If the estimates of the unknown parameters can be expressed explicitly as functions of the observations $\{\mathbf{x}_1, \ldots, \mathbf{x}_n\}$, as is the case for a single normal distribution, then it might be possible to separate out the contribution due to the nth observation, $\mathbf{x}_n$. This permits an inductive sequential formulation of the estimates. From Section 3.3.1 we have

$$\hat{\boldsymbol{\mu}}_n = \hat{\boldsymbol{\mu}}_{n-1} \cdot \frac{n-1}{n} + \mathbf{x}_n \cdot \frac{1}{n}$$

and

$$\hat{\boldsymbol{\Sigma}}_n = \left[\hat{\boldsymbol{\Sigma}}_{n-1} + \frac{(\overline{\mathbf{x}}_{n-1} - \mathbf{x}_n)(\overline{\mathbf{x}}_{n-1} - \mathbf{x}_n)'}{n+1} \right] \frac{n}{n+1}$$

Often, however, such an immediate solution will not be possible and we must resort to rather more complicated approaches. To introduce these let us rewrite the sequential estimate of the mean as

$$\hat{\boldsymbol{\mu}}_n = \hat{\boldsymbol{\mu}}_{n-1} - \frac{1}{n}(\boldsymbol{\mu}_{n-1} - \mathbf{x}_n)$$

Now consider the following theorem:

Theorem (Robbins and Monro, 1951)

Let $g(\theta)$ be an observation, subject to measurement error, of a function $f(\theta)$ and let the measurement error have zero mean and finite variance so that

$$E[g(\theta)] = f(\theta)$$

and

$$E[(g(\theta) - f(\theta))^2] \leqslant \alpha^2 < \infty$$

Let θ_1 be an arbitrary initial estimate of θ^*, the root of $f(\theta)$, with $E(\theta_1^2) < \infty$, and define a sequence of random variables θ_n by

$$\theta_{n+1} = \theta_n - a_n g(\theta_n, x_n) \tag{1}$$

with the x_i independently and identically distributed. Assume further that

$$c_1(\theta^* - \theta)^2 \leqslant f(\theta)(\theta^* - \theta) \leqslant c_2(\theta^* - \theta)^2$$

where $0 < c_1 \leqslant c_2 < \infty$.

Then if the sequence $\{a_n\}$ converges to zero so that $\sum_{n=1}^{\infty} a_n = \infty$ and $\sum_{n=1}^{\infty} a_n^2 < \infty$ the sequence $\{\theta_n\}$ defined in (1) converges to θ^*. ■

We shall not prove this theorem, referring the interested reader to Robbins and Monro (1951), but it is worthwhile taking a closer look at some of the conditions necessary for convergence. The zero mean and finite variance of the error distribution are hardly extreme assumptions and they are required in many statistical models. Note that no assumption has been made about the form of this distribution. The bounds on $f(\theta)(\theta^* - \theta)$ merely imply that $f(\theta)$ lies between two straight lines with slopes c_1 and c_2 intersecting at $\theta = \theta^*$. One implication of this is that $f(\theta)$ has a unique root. The conditions on the a_n prevent endless oscillation of θ_n about θ^* and mean that $\{\theta_n\}$ is not constrained to converge to a limit bounded away from θ^*.

Now, if in this theorem we put $\theta_n = \hat{\mu}_n$, $a_n = 1/n$, and $g(\theta_n, x_n) = \hat{\mu}_n - x_n$ we have expressed the adaptive algorithm for estimating the mean in the Robbins–Monro form. Thus, this sequential estimate of the mean is a special case of a more general procedure.

The simple and convenient adaptive scheme expressed in the above theorem has been extended by several authors. In particular, Blum (1954) has extended it to the multivariate case, Kiefer and Wolfowitz (1952) introduced a modification permitting the estimation of extrema, rather than roots, of a function subject to measurement error, and Dvoretzky (1956) gave a more general form of which both the Robbins–Monro and Kiefer–Wolfowitz algorithms are special cases. However, rather than considering these extensions we shall consider modifications which are more immediate to our aim of parameter estimation.

One can describe the Robbins–Monro algorithm as being a sequential method for estimating the root of a regression function $E[f(\theta^*)] = 0$. Thus, if we can reformulate any particular parameter estimation method into a problem of finding the root of a regression equation we can apply a stochastic approximation algorithm.

In Section 3.4 we pointed out a relationship between the maximum likelihood and distance minimization methods of parameter estimation. We repeat this here with slightly different emphasis—again eschewing mathematical rigour in favour of ease of exposition.

The maximum likelihood solution, $\hat{\theta}$, is a root of

$$\frac{\partial}{\partial \theta}\left[\prod_{i=1}^{n} p(x_i \mid \theta)\right] = 0$$

or, equivalently, of

$$-\frac{1}{n}\frac{\partial}{\partial \theta}\left[\sum_{i=1}^{n} \ln p(x_i \mid \theta)\right] = 0$$

In the limit, as n tends to infinity, we have

$$\lim_{n\to\infty}\frac{1}{n}\sum_{i=1}^{n}\left[-\frac{\partial}{\partial \theta}\ln p(x_i \mid \theta)\right] = E\left[-\frac{\partial}{\partial \theta}\ln p(x \mid \theta)\right]$$

The maximum likelihood solution is thus asymptotically equivalent to finding the root of the regression equation

$$E\left[-\frac{\partial}{\partial \theta}\ln p(x \mid \theta)\right] = 0$$

Substituting this regression function into the Robbins–Monro algorithm yields

$$\theta_{n+1} = \theta_n + a_n \frac{\partial}{\partial \theta}\ln p(x_n \mid \theta)\big|_{\theta_n}$$

We now have a sequential estimation procedure which leads asymptotically to the maximum likelihood solution. Note also the extreme simplicity of the algorithm—the normal equations are often difficult to solve.

As an illustration of this let us take a one-dimensional normal mixture. Thus

$$p(x \mid \boldsymbol{\theta}) = \sum_{i=1}^{c} w_i f(x \mid \mu_i, \sigma_i^2)$$

where $\boldsymbol{\theta} = (\mathbf{w}, \boldsymbol{\mu}, \boldsymbol{\sigma}^2)$. For convenience, following Young and Coraluppi (1970), rather than using σ_i we use $q_i = 1/\sigma_i$. Then the above algorithm yields

$$w_i(n+1) = w_i(n) + a_n \frac{\partial}{\partial w_i} \ln p(x_n | \boldsymbol{\theta}_n), \qquad i = 1, \ldots, c-1$$

$$q_i(n+1) = q_i(n) + a_n \frac{\partial}{\partial q_i} \ln p(x_n | \boldsymbol{\theta}_n), \qquad i = 1, \ldots, c$$

and

$$\mu_i(n+1) = \mu_i(n) + a_n \frac{\partial}{\partial \mu_i} \ln p(x_n | \boldsymbol{\theta}_n), \qquad i = 1, \ldots, c$$

Since the form of $p(x | \boldsymbol{\theta})$ is known we can evaluate the derivatives explicitly.

We have already commented about the problems of the likelihood function for normal mixtures. To alleviate these problems as well as to satisfy the constraints on the weights w_i, Young and Coraluppi have modified this algorithm as follows. They begin by defining parameters d_i with $d_1 = 0$ and $d_i = \sum_{j=1}^{i-1} w_j$, for $i = 2, \ldots, c$. This permits the constraints on the w_i to be rewritten as

$$0 \leq w_i \leq 1 - d_i = 1 - \sum_{j=1}^{i-1} w_j, \qquad j = 1, \ldots, c-1$$

and

$$w_c = 1 - d_c = 1 - \sum_{j=1}^{c-1} w_j$$

They then define

$$\gamma_i = \frac{2w_i - 1 + d_i}{w_i(1 - d_i - w_i)}, \qquad 0 \leq w_i \leq 1 - d_i$$

so

$$w_i = \frac{1}{2\gamma_i}\left[(1 - d_i)\gamma_i - 2 + \sqrt{(1 - d_i)^2\gamma_i^2 + 4}\right]$$

and

$$r_i = q - 1/q_i, \qquad 0 < q_i < \infty$$

(so $q_i = \frac{1}{2}(r_i + \sqrt{r_i^2 + 4})$). These transformations yield the new algorithm

$$\gamma_i(n+1) = \gamma_i(n) + a_i\left[\sum_{j=i}^{c-1} \frac{\partial w_j(n)}{\partial \gamma_i} \cdot \frac{\partial}{\partial w_j} \ln p(x_n | \boldsymbol{\theta}_n)\right]$$

$$r_i(n+1) = r_i(n) + a_i\left[\frac{\partial q_i(n)}{\partial r_i} \cdot \frac{\partial}{\partial q_i} \ln p(x_n | \boldsymbol{\theta}_n)\right]$$

and

$$\mu_i(n+1) = \mu_i(n) + a_i \frac{\partial}{\partial \mu_i} \ln p(x_n|\boldsymbol{\theta}_n)$$

The derivatives can again be evaluated explicitly as

$$\frac{\partial q_i}{\partial r_i} = \frac{1}{2}\left[1 + \frac{r_i}{\sqrt{r_i^2+4}}\right]$$

$$\frac{\partial w_i}{\partial \gamma_i} = \frac{1}{\gamma_i^2}\left[1 - \frac{2}{\sqrt{(1-d_i)^2\gamma_i^2+4}}\right]$$

$$\frac{\partial w_i}{\partial \gamma_i} = -\frac{1}{2}\sum_{k=j}^{i-1} \frac{\partial w_k}{\partial \gamma_j}\left[1 + \frac{(1-d_i)\gamma_i}{\sqrt{(1-d_i)^2\gamma_i^2+4}}\right], \qquad j<i<c$$

One could devise similar schemes for multivariate normal mixtures, though as yet there appears to be no published work on this, or one could use this univariate normal mixture algorithm as a basis for a more general multivariate approach. As Young and Calvert (1974) observe, the marginals of a multivariate normal mixture are also normal. Thus, the above algorithm could be applied on each of the marginals and the within-component covariances estimated subsequently.

A class of sequential estimation algorithms, the members of which can often be described in stochastic approximation terms, is that of 'decision directed' methods. These use the parameter estimates at stage n to classify an incoming point $\mathbf{x}_{n+1}$ as belonging to component j (say) and then update the parameters of this component using $\mathbf{x}_{n+1}$. This bootstrap approach has the advantage of being computationally very simple but has the disadvantages that the estimators often converge not to $\boldsymbol{\theta}$ but to a value near $\boldsymbol{\theta}$ and that a phenomenon called 'runaway' (see Davisson and Schwartz, 1970) can occur. An example of the first disadvantage is given by two overlapping normal components (Figure 3.5). Parameters of component 2 are in fact estimated from a sample drawn from a superposition of two truncated normal distributions (the shaded region) rather than from a single non-truncated distribution. When the problem can be isolated explicitly in this way it can often be adjusted for. The k-means cluster analysis algorithm, when updating occurs after addition of each new point, is an example of a decision directed scheme. Runaway simply describes the occurrence of a few misclassifications leading to more, which in turn lead to more, and so on until the result is very different from the true parameters. Young and Farjo (1972) have considered a decision directed scheme as a method of stochastic approximation.

The above discussion demonstrates that the method of stochastic approximation can provide a general framework for many sequential methods. Apart from this inherent generality it also has the advantage that, once a scheme has been rewritten into a form matching one of the theorems (such as the

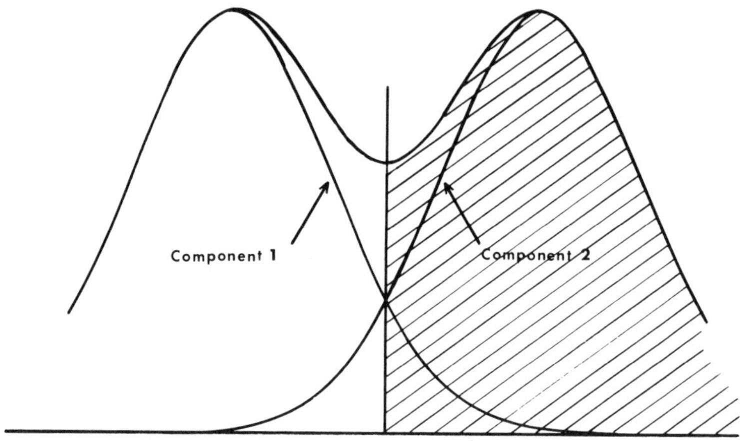

Figure 3.5 Bias in decision directed estimation of a two-component mixture

regression function root-finding formulation of Robbins and Monro), it is usually easy to see if the necessary conditions for convergence are satisfied.

For any iterative or sequential method the question of rate of convergence is important. It is useless having a method guaranteed to converge to the correct solution if such convergence is so slow that a prohibitive number of steps are required. Clearly, for a stochastic approximation algorithm the convergence rate will depend on the sequence $\{a_i\}$. For the Robbins–Monro algorithm with small i it can be shown that a near optimal sequence is given by

$$a_i = \frac{c_1 E(\theta_1 - \theta^*)^2}{\alpha^2 + i c_1^2 E(\theta_1 - \theta^*)^2}$$

(notation as in the statement of Robbins and Monro's theorem). Dvoretzky (1956) has studied this sequence. For large i we can treat the regression function as linear with slope $c \geqslant c_1$ and then an optimum sequence is

$$a_i = 1/ci$$

Sacks (1958) has studied asymptotic distributions of stochastic approximation algorithms.

3.6 SUMMARY

Although advantages can result if use is made of any information known about the forms of the class-conditional pdfs, incorrect or insufficiently close assumptions could lead to an increased error rate. This means that care should be taken when considering a parametric form. One can also combine parametric and non-parametric pdf estimators to yield a two-stage classifier which which has something of the advantages of both types. Thus, a relatively inaccurate classifier based on parameterized pdfs in regions far from the decision surface (i.e. in regions where the difference between the estimated class-conditional pdfs exceeds some threshold) can be supplemented by a more accurate non-parametric approach in doubtful regions. However, one should note that in doubtful regions the classification will be in error a large proportion of the time in any case.

For various reasons the normal distribution and its generalization to normal mixtures are popular families of class-conditional pdfs.

Although the description of the estimation techniques has been aided by the grouping into three categories, a more revealing classification may be into the two groups:

(i) maximum likelihood and distance minimization, and
(ii) Bayes methods.

In each case our aim is to estimate the pdf for each class, given by $p(\mathbf{x}|\boldsymbol{\theta}_i)$. In the first case we use the sample to provide an estimate $\hat{\boldsymbol{\theta}}_i$ of $\boldsymbol{\theta}_i$ to give

$$p(\mathbf{x}|\mathbf{x}_1, \ldots, \mathbf{x}_n) = p(\mathbf{x}|\hat{\boldsymbol{\theta}}_i)$$

In the second case we take a weighted sum of the $p(\mathbf{x}|\boldsymbol{\theta}_i)$ with weights given by the relative probability of a particular $\boldsymbol{\theta}_i$-value having given rise to the observed data. That is

$$p(\mathbf{x}|\mathbf{x}_1, \ldots, \mathbf{x}_n) = \int p(\mathbf{x}|\boldsymbol{\theta}_i)r(\boldsymbol{\theta}_i|\mathbf{x}_1, \ldots, \mathbf{x}_n)\, d\boldsymbol{\theta}_i$$

The second approach can be viewed as an attempt to make allowance for the variability inherent in $\hat{\boldsymbol{\theta}}_i$, something which is lacking in the first approach. On the other hand, the second method does require a prior distribution $r(\boldsymbol{\theta}_i)$ which, while well investigated for the normal cases illustrated, will be more difficult for more complicated families of pdfs. Even given a suitable prior, unless the choice of class-conditional pdfs admits of reproducing densities (that is, unless there exist sufficient statistics) the final evaluation of $p(\mathbf{x}|\mathbf{x}_1, \ldots, \mathbf{x}_n)$ may need to be via two multivariate numerical integrations.

Sequential methods are important for several reasons and in different areas of pattern recognition. The method of stochastic approximation provides a unified framework for many sequential methods—in particular, as Chien and Fu (1967) show, Bayesian algorithms can be seen to be members of this class.

The problem of parameter estimation is a classical one in statistics and has been dealt with in numerous textbooks. In particular we refer the reader to Kendall and Stuart (1961, vol. 2), Wilks (1963), and Van Der Waerden (1969). More details of Bayesian methods can be found in De Groot (1970), Aitchison and Dunsmore (1975), and Raiffa and Schlaifer (1961). For a treatment of the various ways to estimate parameters of mixture distributions, see Everitt and Hand (1981). An excellent introduction to stochastic approximation methods is provided by Young and Calvert (1974, ch. 5), an outline which influenced this presentation.

EXERCISES

3.1 By referring to the proof in Section 3.3.1, investigate the effect of requiring $\mathbf{\Sigma}_i \propto \mathbf{\Sigma}_1$ (for all $i = 2, \ldots, c$) on the unboundedness of the likelihood function for normal mixtures.

3.2 Consider the following two-component bivariate normal mixtures with parameters $\boldsymbol{\mu}_1$, $\boldsymbol{\mu}_2$, $\mathbf{\Sigma}_1$, $\mathbf{\Sigma}_2$, and p:

(a) $\mathbf{\Sigma}_1 = \alpha\mathbf{\Sigma}_2$, α a constant, and

(b) $\mathbf{\Sigma}_1 = \mathbf{D}_1$, $\mathbf{\Sigma}_2 = \mathbf{D}_2$, where the $\mathbf{D}_i$ are diagonal.

Then mixture (a) has nine parameters ($\boldsymbol{\mu}_1$, $\boldsymbol{\mu}_2$, $\mathbf{\Sigma}_1$, α, and p) and mixture (b) also has nine parameters ($\boldsymbol{\mu}_1$, $\boldsymbol{\mu}_2$, $\mathbf{D}_1$, $\mathbf{D}_2$, and p).

Through simulation studies and by applying (a) and (b) to real data sets, investigate whether one family can be expected to perform better than the other.

3.3 The underlying population distribution of a d-dimensional sample is estimated by two mixture models:

(a) a c class mixture with each component having identity variance covariance matrix, and

(b) a single normal distribution.

Model (a) requires $cd + (c - 1) + d$ parameters.

Model (b) requires $d + d(d + 1)/2$ parameters.

If we choose $c = 1/(d + 1) + d/2$ then (a) and (b) have the same number of parameters. Using the iterative maximum likelihood algorithm of Section 3.3.1, compare the models.

How does sample size affect the comparison?

3.4 Given two samples known to come from classes with normal population distributions with equal priors, are there advantages to be gained by using the Bayesian (predictive) approach of Section 3.3.3?

3.5 Through simulation studies applied to data from two multivariate normal populations with equal variance–covariance matrices, compare the performance of the Bayesian method of Section 3.3.3 with the error correction algorithms of Chapter 4.

How do the classification times compare?

How do the results depend on design set size?

3.6 The parametric methods outlined in this chapter have both advantages and disadvantages. Examples of the former are the facts that once the parameters have been estimated the design set can be discarded and that classifications in subspaces can easily be made (for example, to classify vectors with missing components). An obvious example of the latter is that assumptions about distributional form must be made. These methods have their origin primarily in the classical statistical side of the subject. The non-parametric methods outlined in Chapter 2, on the other hand, have their origins primarily in pattern recognition work and are relatively recent developments. Suggest reasons for these differences. In the light of these reasons, how do you see the subject developing in the future?

CHAPTER 4

Linear Discriminant Functions

4.1 INTRODUCTION

At the end of Section 1.3 we defined a discriminant function for the two-class case as a general function $h(\mathbf{x})$ leading to the classification rule

$$h(\mathbf{x}) \gtrless k \Rightarrow \mathbf{x} \in \begin{cases} \Omega_1 \\ \Omega_2 \end{cases} \tag{1}$$

where k is a constant. Also in Chapter 1 we demonstrated that $h(\mathbf{x})$ of the form

$$h(\mathbf{x}) = p(\mathbf{x} \mid \omega_1)/p(\mathbf{x} \mid \omega_2) \tag{2}$$

gave optimum solutions using the minimum error and minimum risk criteria. Since the class-conditional pdfs, $p(\mathbf{x} \mid \omega_i)$, are unknown, implementing this $h(\mathbf{x})$ required estimating the $p(\mathbf{x} \mid \omega_i)$ from the design set. In Chapter 3 we expedited this estimation by assuming that the $p(\mathbf{x} \mid \omega_i)$ came from known families of distributions, leaving only parameters to be estimated. Here, instead of making assumptions about the $p(\mathbf{x} \mid \omega_i)$, we make assumptions about the discriminant functions directly. Specifically, we assume that these functions have a known form with unknown parameters. However, since our aim is simply to identify regions Ω_i, the absolute values of $h(\mathbf{x})$ are irrelevant provided (1) is satisfied. More formally, any function $g = g(\mathbf{x})$ will serve as a suitable discriminant function if it satisfies

$$g(\mathbf{x}) \gtrless k_g \Leftrightarrow h(\mathbf{x}) \gtrless k \tag{3}$$

where k_g is a suitably chosen constant. (Compare the outlines in Chapter 2 of the potential function method and the condensed nearest-neighbour method.) We can make use of this freedom of choice in g by selecting a form which is computationally convenient. In particular, advantages will result if we choose g to be linear in the components of $\mathbf{x}$. The design problem then becomes one of estimating coefficients of the x_i to optimize the classification rule

$$g(\mathbf{x}) \gtrless k_g \Rightarrow \mathbf{x} \in \begin{cases} \Omega_1 \\ \Omega_2 \end{cases}$$

While computationally convenient, it might be felt that choosing $g(\mathbf{x})$ to be linear must impose severe restrictions on the resulting classifier—after all, if $g(\mathbf{x})$ is linear then the decision surface $h(\mathbf{x}) = k$ is linear. However, the situation is not quite as bad as all that. First consider the practically and theoretically important special case of normal class-conditional pdfs with equal variance–covariance matrices. If we define

$$g(\mathbf{x}) = \ln(h(\mathbf{x}))$$

with h as in (2) then, since ln is a monotonic function, it is obvious that

$$g(\mathbf{x}) \gtrless k_g \Leftrightarrow h(\mathbf{x}) \gtrless k$$

(here $k_g = \ln(k)$). Moreover, as shown in Section 4.8, for this special case $\ln h(\mathbf{x})$ is a linear function.

Secondly, our discussion has been in terms of linearity in components of $\mathbf{x}$. However, there is no reason why we should not increase the dimensionality of the space by adding to $\mathbf{x}$ extra higher order terms. More generally, we can replace $\mathbf{x}$ in the whole of the preceding discussion by

$$\boldsymbol{\phi}(\mathbf{x}) = (\phi_1(\mathbf{x}), \ldots, \phi_D(\mathbf{x}))$$

a vector of functions of $\mathbf{x}$. (In the terminology of Section 1.2, we can call the x_i measurements and the ϕ_i, *features*.) Then, instead of requiring g to be linear in the x_i we require it to be linear in the ϕ_i, i.e. g is a linear function *of functions* of the x_i. An obvious example of this sort of approach is generalizing g from a linear function

$$g(\mathbf{x}) = \sum_i a_i x_i$$

to being a polynomial function

$$g(\mathbf{x}) = \sum_i a_i x_i + \sum_{i \leq j} \sum a_{ij} x_i x_j$$

For obvious reasons the discriminant functions employed in these kinds of approaches are termed *generalized linear discriminant functions*.

One can also extend the range of applicability of linear functions by using several simultaneously—the *piecewise linear* functions mentioned in Section 4.8.

Having suggested how the apparently limited range of linear discriminant functions can be extended, it is perhaps as well to add a cautionary note. The Bayesian decision rule is optimal and unless one's chosen form for the discriminant function leads to the same (or a sufficiently similar) decision surface, a poor classifier could result. Whereas in principle one could increase the number of parameters and hence the rule's flexibility *ad infinitum*, in practice this is not feasible. First there are the problems associated with only having a finite design set discussed in Chapter 6, and secondly the further one gets away from the simple linear function so the more one loses the computational advantages of such functions. It should also be noted that in high dimensional spaces it might not be feasible to use anything above simple first order terms in the x_i.

In the remainder of this chapter we shall assume discriminant functions linear in the x_i. More general cases follow immediately by substituting ϕ_i for x_i.

4.2 GENERAL IDEAS

The exposition will be considerably simplified by restricting it to the two-class case. This is the most common special case and is of great importance. At the end of this chapter we propose several ways of extending the methods to more than two classes.

We begin by trying to find a linear function

$$g(\mathbf{x}) = \mathbf{v}'\mathbf{x} + v_0$$

$$\mathbf{v}'\mathbf{x} + v_0 \gtrless 0 \Rightarrow \mathbf{x} \in \begin{cases} \Omega_1 \\ \Omega_2 \end{cases} \tag{1}$$

We can simplify the subsequent mathematics by defining

$$\mathbf{z}' = (1, \mathbf{x}') \quad \text{and} \quad \mathbf{w}' = (v_0, \mathbf{v}')$$

so that (1) becomes

$$\mathbf{w}'\mathbf{z} \gtrless 0 \Rightarrow \mathbf{x} \in \begin{cases} \Omega_1 \\ \Omega_2 \end{cases}$$

$\mathbf{w}$ is called the *weight* vector and $\mathbf{z} = (1, \mathbf{x}')'$ is the augmented observation vector.

The weight vector $\mathbf{w}$ must be estimated from the two sets of samples comprising the design set, and in view of the decision rule above a sensible estimate would be one for which

(a) $\mathbf{w}'\mathbf{z}_i > 0$, whenever $\mathbf{x}_i$ is in the sample from class ω_1, and
(b) $\mathbf{w}'\mathbf{z}_i < 0$, whenever $\mathbf{x}_i$ is in the sample from class ω_2.

Here $\{\mathbf{z}_i = (1, \mathbf{x}_i')'\}$ is the set of augmented design set vectors. Things will be simplified still further (again without loss of generality) if we define $\mathbf{y}_i$ by

(a) $\mathbf{y}_i = \mathbf{z}_i = (1, \mathbf{x}_i')'$ for $\mathbf{x}_i$ in the sample from class ω_1 and
(b) $\mathbf{y}_i = -\mathbf{z}_i = (-1, -\mathbf{x}_i')'$ for $\mathbf{x}_i$ in the sample from class ω_2, introducing the minus signs.

Now the sensible estimate of $\mathbf{w}$ must satisfy

$$\mathbf{w}'\mathbf{y}_i > 0, \quad \text{for all } \mathbf{y}_i \text{ corresponding to } \mathbf{x}_i \text{ in the design set.} \tag{2}$$

(For convenience, in what follows we shall talk of 'all $\mathbf{y}_i$ in the design set'. This simply means all $\mathbf{y}_i = \pm(1, \mathbf{x}_i')'$, where $\mathbf{x}_i$ is in the design set.)

It is clear that the decision surface estimated from this discriminant function is $\mathbf{w}'\mathbf{y} = 0$, which is, of course, a hyperplane, so that unless the two samples can be separated by such a surface (i.e. unless they are *linearly separable*) we will not be able to find a $\mathbf{w}$ such that (2) is satisfied. Our ideal is thus modified to make $\mathbf{w}'\mathbf{y}_i > 0$ for as many of the sample points as possible. For obvious reasons this criterion is known as the *misclassification criterion*.

This is a convenient place to remind the reader of a point raised in Section 1.4, namely that the decision surface which minimizes the design set misclassification rate (or, indeed, any other criterion) is unlikely to be the Bayes optimal decision surface—even if the latter is linear. This is because the estimate of $\mathbf{w}$ is based only on a sample from the underlying populations and not on perfect descriptions of these populations. Obviously the larger the sample the better, and for reasonably sized samples one would expect the methods outlined below to yield similar results.

Unfortunately the number of misclassifications is a difficult criterion to minimize. One of the simplest extensions is the *perceptron criterion*, defined by

$$C_1(\mathbf{w}) = \sum_M (-\mathbf{w}'\mathbf{y}_i) \tag{3}$$

where the summation is over the set $M = \{\mathbf{y}_i \text{ such that } \mathbf{w}'\mathbf{y}_i < 0\}$. The name comes from a brain model devised by Rosenblatt (1962). From the definition it is clear that only points which are misclassified contribute to C_1, and they do so to an extent proportional to their distance from the hyperplane $\mathbf{w}'\mathbf{y} = 0$, the decision surface (whereas in the misclassification criterion each misclassified point contributes an equal amount). The next two sections outline ways to minimize this criterion.

4.3 LINEAR PROGRAMMING AND THE PERCEPTRON CRITERION

Linear programming deals with problems in which a linear function is to be maximized or minimized subject to specified linear constraints and with the additional conditions that all variables must take non-negative values. This is not the place to describe the methods of linear programming, and readers unfamiliar with them are referred to appropriate texts (for example, Vajda, 1970, and Dantzig, 1963). Here we simply observe that the perceptron criterion is a piecewise linear function of $\mathbf{w}$ and show how the perceptron criterion minimization problem can be reformulated so that linear programming may be applied.

First note that if each component of $\mathbf{w}$ tends to zero while the direction $\mathbf{w}/|\mathbf{w}|$ remains constant, then $C_1(\mathbf{w})$ in (3) of Section 4.2 tends to zero. Rounding error in the computations means that in practice the 'solution' $\mathbf{w} = \mathbf{0}$ could result. To avoid this we must impose some additional constraint.

We could normalize $\mathbf{w}$, requiring $\mathbf{w}'\mathbf{w} = 1$, but this would sacrifice the piecewise linear nature of $C_1(\mathbf{w})$. So instead let us define a set

$$\{b_i\}, \qquad i = 1, \ldots, n, \text{ each } b_i > 0$$

and replace the requirement

$$\mathbf{w}'\mathbf{y}_i > 0, \quad \text{for all } y_i \text{ in the design set}$$

by

$$\mathbf{w}'\mathbf{y}_i > b_i, \quad \text{for all } y_i \text{ in the design set}$$

The first condition is certainly true if the second is. By analogy with (3) above we now try to minimize

$$C_2(\mathbf{w}) = \sum_{M^*} (b_i - \mathbf{w}'\mathbf{y}_i)$$

where $M^* = \{\mathbf{y}_i \text{ such that } b_i - \mathbf{w}'\mathbf{y}_i \geq 0\}$. C_2 does not tend to zero if the components of $\mathbf{w}$ tend to zero. Note that it is no longer only the misclassified points which contribute to C_2, but also those correctly classified points which lie closer than distance $b_i/\sqrt{\mathbf{w}'\mathbf{w}}$ to the hyperplane $\mathbf{w}'\mathbf{y} = 0$. Thus, a safety margin has been built in; by minimizing C_2 we are trying to find a decision surface which not only classifies all points correctly but which moreover has no points lying closer than $b_i/\sqrt{\mathbf{w}'\mathbf{w}}$ to it. (But note that the importance of this safety margin—its width—does depend on the length of $\mathbf{w}$.) The idea of a safety margin will recur later.

Since C_2 is *piecewise* linear we cannot apply linear programming methods immediately so we introduce n (the total sample size) 'artificial variables', a_i, which minimize

$$Z = \sum_{i=1}^{n} a_i$$

subject to

$$a_i \geq 0$$
$$a_i \geq b_i - \mathbf{w}'\mathbf{y}_i$$

Since the minimum of any a_i is $\max(0, b_i - \mathbf{w}'\mathbf{y}_i)$ we have

$$\min_{\mathbf{a},\mathbf{w}} Z = \min_{\mathbf{w}} C_2$$

In order to couch this in standard linear programming format we should also impose the condition $\mathbf{w} \geq \mathbf{0}$, but this is too restrictive for us. However, we can satisfy this if we define $|\mathbf{w}|$ = the vector $\mathbf{w}$ with all negative components replaced by positive ones. Then we can define

$$\mathbf{w}^+ = \tfrac{1}{2}(|\mathbf{w}| + \mathbf{w})$$

and

$$\mathbf{w}^- = \tfrac{1}{2}(|\mathbf{w}| - \mathbf{w})$$

Both $\mathbf{w}^+$ and $\mathbf{w}^-$ have all components ≥ 0 and

$$\mathbf{w} = \mathbf{w}^+ - \mathbf{w}^-$$

The perceptron criterion minimization problem has now been reformulated as

$$\min_{\mathbf{a},\mathbf{w}} \sum_{i=1}^{n} a_i$$

subject to

$$\begin{aligned} a_i &\geq 0 \\ \mathbf{w}^+ &\geq \mathbf{0} \\ \mathbf{w}^- &\geq \mathbf{0} \\ a_i &\geq b_i - \mathbf{w}^{+\prime}\mathbf{y}_i + \mathbf{w}^{-\prime}\mathbf{y}_i \end{aligned}$$

Standard linear programming computer packages can be applied in tackling this new formulation of the problem.

We illustrate using the data of Table 4.1. This is from Margolese (1970) who was studying the hypothesis that diminished androgen/oestrogen ratios predisposed males towards homosexuality. We arbitrarily set the safety margins b_i all equal to 1.0 and used the simplex linear programming algorithm. In general there are n a_i variables, $(d + 1)$ w_j^+ variables, $(d + 1)$ w_j^- variables, and a further n 'slack' variables added by the algorithm to convert the constraints into equations. In our example there are thus a total of $26 + 3 + 3 + 26 = 58$ variables. Since each sample point is associated with

Table 4.1 Values of urinary androsterone and etiocholanolone in healthy heterosexual and homosexual males in mg/24 hours (data from Margolese, 1970)

	Heterosexual			Homosexual	
	Androsterone	Etiocholanolone		Androsterone	Etiocholanolone
1	3.9	1.8	12	2.5	2.1
2	4.0	2.3	13	1.6	1.1
3	3.8	2.3	14	3.9	3.9
4	3.9	2.5	15	3.4	3.6
5	2.9	1.3	16	2.3	2.5
6	3.2	1.7	17	1.6	1.7
7	4.6	3.4	18	2.5	2.9
8	4.3	3.1	19	3.4	4.0
9	3.1	1.8	20	1.6	1.9
10	2.7	1.5	21	4.3	5.3
11	2.3	1.4	22	2.0	2.7
			23	1.8	3.6
			24	2.2	4.1
			25	3.1	5.2
			26	1.3	4.0

a constraint, there are 26 such constraints. (If n is large the number of constraints might make the linear programming formulation impracticable.) We used the NAG subroutine H01ADF on our data. It converged after 29 iterations to the decision surface

$$-6.03 + 5.81x_1 - 4.52x_2 = 0$$

which is line C of Figure 4.1 (x_1 refers to androgen).

4.4 ERROR CORRECTION AND THE PERCEPTRON CRITERION

Since the perceptron criterion is a continuous function it can be minimized by steepest descent methods. We have

$$C_1(\mathbf{w}) = \sum_M (-\mathbf{w}'\mathbf{y}_i)$$

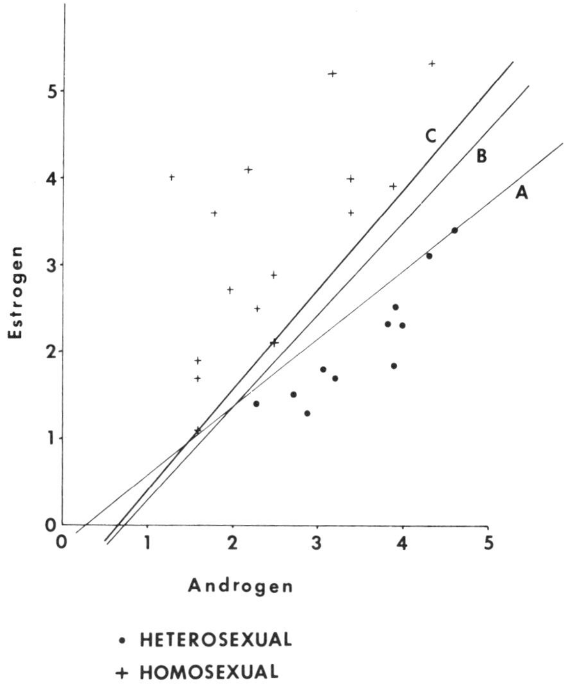

Figure 4.1 Values of urinary androsterone and etiocholanolone in 11 healthy heterosexual and 15 healthy homosexual males (mg/24 hours). (a) Decision surface for error correction algorithm with $\rho_k = 1/k$ and no margin. (b) Decision surface for error correction algorithm with $\rho_k = 1.0$ and margin $b = 1.0$. (c) Linear programming solution. (Margolese, 1970)

where M = {samples, $\mathbf{y}_i$, misclassified by $\mathbf{w}$} so

$$\nabla C_1 = -\sum_M \mathbf{y}_i$$

and

$$\mathbf{w}_{k+1} = \mathbf{w}_k + \rho_k \sum_{M_k} \mathbf{y}_i \tag{1}$$

Here $\mathbf{w}_k$ is the weight vector at the kth iteration and M_k = {samples, $\mathbf{y}_i$, misclassified by $\mathbf{w}_k$}. ρ_k is a function of k which determines step size. This is an example of an error correction procedure since $\mathbf{w}_k$ is only adjusted when some of the $\mathbf{y}_i$ are misclassified.

This steepest descent method for calculating $\mathbf{w}$ has the property that when the sample sets are linearly separable $\mathbf{w}_k$ is guaranteed to converge to a solution. Rather than proving convergence for (1), however, we shall consider a modification

$$\mathbf{w}_{k+1} = \mathbf{w}_k + \rho_k \mathbf{y}_i$$

where $\mathbf{y}_i$ is a single point in the design set which has been misclassified by $\mathbf{w}_k$. This procedure iterates cyclically through the design samples, altering $\mathbf{w}_k$ whenever one is misclassified, rather than considering all of the samples which are misclassified by a particular $\mathbf{w}_k$. This sort of approach can have advantages. For further simplicity we shall only prove convergence for the case $\rho_k = \rho$. Other convergence proofs may be found in Nilsson (1965). Note also that an arbitrary scaling of the $\mathbf{y}_i$ makes no difference, so it is sufficient to prove the result for $\rho = 1$.

Theorem

If the two sample sets are linearly separable and if the updating procedure is defined as

$$\mathbf{w}_{k+1} = \mathbf{w}_k + \mathbf{y}_i$$

whenever a sample element $\mathbf{y}_i$ is misclassified by $\mathbf{w}_k$ during the course of repeated iteration through the sample sets, then $\mathbf{w}_k$ converges to a solution vector $\mathbf{w}$ which linearly separates the two sample sets.

Proof. Since the sample sets are linearly separable we know that there does exist a solution vector $\mathbf{w}$ such that

$$\mathbf{w}'\mathbf{y}_i > 0, \quad \text{for all } \mathbf{y}_i \text{ in the design set}$$

For convenience we shall normalize $\mathbf{w}$ by α where

$$\alpha = \max_B \mathbf{y}_i'\mathbf{y}_i / \min_B \mathbf{w}'\mathbf{y}_i$$

where B is the entire design set. Since $\mathbf{w}'\mathbf{y}_i > 0$ for all $\mathbf{y}_i \in B$, α is positive and this normalization makes no difference to the classification rule.

Now, if we could show that the difference between $\alpha\mathbf{w}$ and $\mathbf{w}_k$ decreases at least by a known positive constant at each step, then (since sensible choices of $\mathbf{w}_1$ lead to a finite value for $(\mathbf{w}_1 - \alpha\mathbf{w})'(\mathbf{w}_1 - \alpha\mathbf{w})$; for example, $\mathbf{w}_1 = \mathbf{0}$) $\mathbf{w}_k$ must converge to $\alpha\mathbf{w}$ in a finite number of steps.

$\mathbf{w}_{k+1}$ is given by

$$\mathbf{w}_{k+1} = \mathbf{w}_k + \mathbf{y}_i$$

so that

$$\mathbf{w}_{k+1} - \alpha\mathbf{w} = \mathbf{w}_k - \alpha\mathbf{w} + \mathbf{y}_i$$

and

$$(\mathbf{w}_{k+1} - \alpha\mathbf{w})'(\mathbf{w}_{k+1} - \alpha\mathbf{w}) = (\mathbf{w}_k - \alpha\mathbf{w})'(\mathbf{w}_k - \alpha\mathbf{w}) + \mathbf{y}_i'\mathbf{y}_i + 2(\mathbf{w}_k - \alpha\mathbf{w})'\mathbf{y}_i$$

Since $\mathbf{y}_i$ is misclassified by $\mathbf{w}_k$ we know that $\mathbf{w}_k'\mathbf{y}_i \leqslant 0$ so

$$(\mathbf{w}_{k+1} - \alpha\mathbf{w})'(\mathbf{w}_{k+1} - \alpha\mathbf{w}) \leqslant (\mathbf{w}_k - \alpha\mathbf{w})'(\mathbf{w}_k - \alpha\mathbf{w}) + \mathbf{y}_i'\mathbf{y}_i - 2\alpha\mathbf{w}'\mathbf{y}_i$$

Hence

$$(\mathbf{w}_{k+1} - \alpha\mathbf{w})'(\mathbf{w}_{k+1} - \alpha\mathbf{w}) \leqslant (\mathbf{w}_k - \alpha\mathbf{w})'(\mathbf{w}_k - \alpha\mathbf{w}) + \max_B \mathbf{y}_i'\mathbf{y}_i - 2\alpha \min_B \mathbf{w}'\mathbf{y}_i$$

Now, by definition of α

$$(\mathbf{w}_{k+1} - \alpha\mathbf{w})'(\mathbf{w}_{k+1} - \alpha\mathbf{w}) \leqslant (\mathbf{w}_k - \alpha\mathbf{w})'(\mathbf{w}_k - \alpha\mathbf{w}) - \max_B \mathbf{y}_i'\mathbf{y}_i$$

Since $\max_B \mathbf{y}_i'\mathbf{y}_i$ is positive, we have completed the proof. ■

In fact the above theorem shows that, starting from $\mathbf{w}_1$, we shall arrive at a solution after at most

$$\beta = (\mathbf{w}_1 - \alpha\mathbf{w})'(\mathbf{w}_1 - \alpha\mathbf{w})/\max_B \mathbf{y}_i'\mathbf{y}$$

corrections.

Up to now we have interpreted the decision surface

$$\mathbf{w}'\mathbf{y} = 0$$

as a hyperplane in $\mathbf{y}$-space (i.e. the space with components of $\mathbf{y}$ as axes). $\mathbf{w}$ is the normal to that hyperplane. However, there is no reason why we should not invert the representation and regard $\mathbf{y}$ as the normal to the hyperplane in $\mathbf{w}$ or *weight space* (the space with components of $\mathbf{w}$ as axes). This interpretation can be a convenient intuitive aid. In the weight space $\mathbf{w}$ is a point and each $\mathbf{y}$ defines a hyperplane.

Figure 4.2 gives a two-dimensional example of a weight space representation for a problem with four sample points. The corresponding hyperplanes are indicated by numbers. The arrows show the positive sides of the hyperplanes and the solution must lie in the shaded region, which is on the positive side of every hyperplane. This region is called the *solution region*. By following a path of $\mathbf{w}_k$ for successive corrections $\mathbf{w}_{k+1} = \mathbf{w}_k + \mathbf{y}_i$ in this

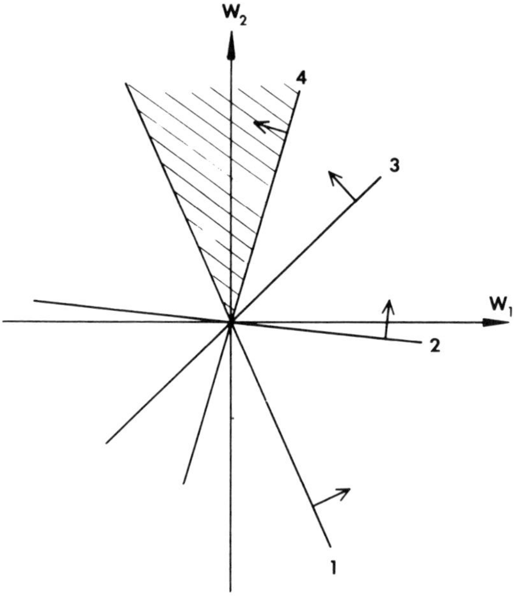

Figure 4.2 The weight–space representation

figure it is easy to see why the procedure must eventually produce a $\mathbf{w}_k$ in the shaded region.

We proved convergence of a simple version of the updating algorithm but it can be shown that if the samples are linearly separable then

$$\mathbf{w}_{k+1} = \mathbf{w}_k + \rho_k \mathbf{y}$$

converges if

(i) $$\rho_k \geqslant 0$$

(ii) $$\sum_{k=1}^{\infty} \rho_k = \infty$$

(iii) $$\lim_{m \to \infty} \frac{\sum_{k=1}^{m} \rho_k^2}{\left(\sum_{k=1}^{m} \rho_k \right)^2} = 0$$

If the samples are not linearly separable then a method with ρ_k constant will continue to leap erratically whereas one with $\rho_k \to 0$ (a common choice is $\rho_k = 1/k$) will decrease the contribution from misclassified points at each iteration. Unfortunately, condition (ii) above, which is needed to guarantee a solution if one exists, prevents us from choosing a convergent series for ρ_k.

The perceptron criterion is of considerable historical interest and many improvements and modifications of it exist. One of its disadvantages is that a solution $\mathbf{w}$ can be found which lies near the edge of the solution region. In practical terms this simply means that although the hyperplane $\mathbf{w}'\mathbf{y} = 0$ separates the two sample sets some of the points lie near to the hyperplane. This means that the hyperplane might not generalize well—its performance on new vectors could be improved. One way to achieve this improvement is by defining a margin $b > 0$ and requiring $\mathbf{w}'\mathbf{y}_i \geqslant b$. We have already used this idea in Section 4.3. The effect of this in weight space is shown in Figure 4.3.

This algorithm was applied to the Margolese (1970) data of Table 4.1 and Figure 4.1. Decision surface A in the figure is the result obtained using $\rho_k = 1/k$ and no margin. The first and last few steps of the iterative process are given in Table 4.2. In all, 15 cycles through the 26 data points were needed, with corrections being made after points had been misclassified 36 times. The starting position was $(w_0 = w_1 = w_2 = 1)$. Decision surface B resulted from the same starting point when ρ_k = constant = 1.0 and a margin of $b = 1.0$ was used. The first and last stages of the estimation are shown in Table 4.3. Eighteen cycles through the data were required, with 42 updates occurring before a solution was achieved. Note the effect of the margin in forcing the decision surface away from its nearest design set point.

Other criteria have been suggested (see, for example, Duda and Hart, 1973, p. 148) and each has different advantages and disadvantages. In the

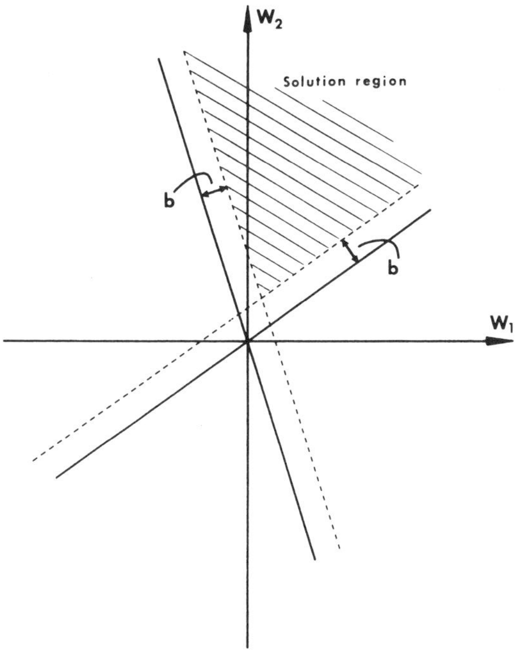

Figure 4.3 Effect of a margin in weight–space

Table 4.2 Error correction algorithm for the perceptron criterion using no margin and $\rho_k = 1/k$

k	Updating point: Number	Updating point: $\mathbf{y}$	ρ_k	$\mathbf{w}$	
—	—	—	—	(1, 1, 1)	$\mathbf{w}_1$
1	12	(−2.5, −2.1, −1)	1	(−1.5, −1.1, 0)	$\mathbf{w}_2$
2	1	(3.9, 1.8, 1)	1/2	(0.5, −0.2, 0.5)	$\mathbf{w}_3$
3	12	(−2.5, −2.1, −1)	1/3	(−0.4, −0.9, 0.2)	$\mathbf{w}_4$
4	1	(3.9, 1.8, 1)	1/4	(0.6, −0.4, 0.4)	$\mathbf{w}_5$
⋮	⋮	⋮	⋮	⋮	⋮
33	13	(−1.6, −1.1, −1)	1/33	(0.7, −0.9, −0.2)	$\mathbf{w}_{34}$
34	7	(4.6, 3.4, 1)	1/34	(0.9, −0.8, −0.1)	$\mathbf{w}_{35}$
35	12	(−2.5, −2.1, −1)	1/35	(0.8, −0.9, −0.1)	$\mathbf{w}_{36}$
36	13	(−1.6, −1.1, −1)	1/36	(0.7, −0.9, −0.2)	$\mathbf{w}_{37}$

Final weight vector $\mathbf{w}_{37}$ = (0.7, −0.9, −0.2).

next section we outline what was perhaps the first formal criterion to be used in discrimination problems, namely that adopted by Fisher in 1936.

4.5 FISHER'S CRITERION

Our problem is one of identifying that linear surface which best discriminates between the two classes. This is formally equivalent to finding that direction (the normal to the above linear surface) along which the two groups are best

Table 4.3 Error correction algorithm for the perceptron criterion using margin b = 1.0 and ρ_k = 1.0 for all k

k	Updating point: Number	Updating point: $\mathbf{y}$	ρ_k	$\mathbf{w}$	
—	—	—	—	(1, 1, 1)	$\mathbf{w}_1$
1	12	(−2.5, −2.1, −1.0)	1.0	(−1.5, −1.1, 0)	$\mathbf{w}_2$
2	1	(3.9, 1.8, 1.0)	1.0	(2.4, 0.7, 1.0)	$\mathbf{w}_3$
3	12	(−2.5, −2.1, −1.0)	1.0	(−0.1, −1.4, 0.0)	$\mathbf{w}_4$
4	1	(3.9, 1.8, 1.0)	1.0	(3.8, 0.4, 1.0)	$\mathbf{w}_5$
⋮	⋮	⋮	⋮	⋮	⋮
39	11	(2.3, 1.4, 1.0)	1.0	(10.3, −6.9, −6.0)	$\mathbf{w}_{40}$
40	12	(−2.5, −2.1, −1.0)	1.0	(7.8, −9.0, −7.0)	$\mathbf{w}_{41}$
41	4	(3.9, 2.5, 1.0)	1.0	(11.7, −6.5, −6.0)	$\mathbf{w}_{42}$
42	12	(−2.5, −2.1, 1.0)	1.0	(9.2, −8.6, −7.0)	$\mathbf{w}_{43}$

Final weight vector $\mathbf{w}_{43}$ = (9.2, −8.6, −7.0).

separated. Fisher defined the separation between two groups in a particular direction as the distance between the means of the two groups standardized for the within-group variance in the specified direction. Figure 4.4 illustrates how important this standardization is. Prior to standardization (the left of the figure) the separation in the x_1 direction, $(\mu_1 - \nu_1)$, seems much greater than the separation in the x_2 direction, $(\mu_2 - \nu_2)$. However, when we standardize so that the variances in the x_1 and x_2 directions are the same (the right-hand side of the figure) it becomes apparent that the separation in direction x_2 is greater. Of course, in our search for the best separating direction we are not constrained to choose one of the coordinate directions. We thus want to find the direction $\mathbf{v}$, such that $(\mathbf{v}'\bar{\mathbf{x}}_1 - \mathbf{v}'\bar{\mathbf{x}}_2)$ is maximized relative to the standard deviation $\sqrt{\mathbf{v}'\mathbf{S}\mathbf{v}}$ in that direction. Here $\bar{\mathbf{x}}_i$ is the sample mean of the design set for class ω_i $(i = 1, 2)$ and $\mathbf{S}$ is the *assumed common* sample variance–covariance matrix. That is, we want to choose $\mathbf{v}$ to maximize the ratio

$$C_3 = \frac{\text{distance between sample means}}{\text{standard deviation within samples}} = \frac{\mathbf{v}'\bar{\mathbf{x}}_1 - \mathbf{v}'\bar{\mathbf{x}}_2}{\sqrt{\mathbf{v}'\mathbf{S}\mathbf{v}}}$$

Differentiating this with respect to $\mathbf{v}$ and equating it to $\mathbf{0}$ yields

$$\bar{\mathbf{x}}_1 - \bar{\mathbf{x}}_2 = \hat{\mathbf{v}}'(\bar{\mathbf{x}}_1 - \bar{\mathbf{x}}_2)\mathbf{S}\hat{\mathbf{v}}/\sqrt{\hat{\mathbf{v}}'\mathbf{S}\hat{\mathbf{v}}}$$

Now, we are solely interested in the direction of $\hat{\mathbf{v}}$, i.e. in the relative magnitude of its components. Their absolute value is irrelevant. Thus, multiplying $\hat{\mathbf{v}}$ by an arbitrary scalar makes no difference. In particular

$$\hat{\mathbf{v}}'(\bar{\mathbf{x}}_1 - \bar{\mathbf{x}}_2)/\sqrt{\hat{\mathbf{v}}'\mathbf{S}\hat{\mathbf{v}}}$$

is a scalar. We thus finally have

$$\hat{\mathbf{v}} \propto \mathbf{S}^{-1}(\bar{\mathbf{x}}_1 - \bar{\mathbf{x}}_2)$$

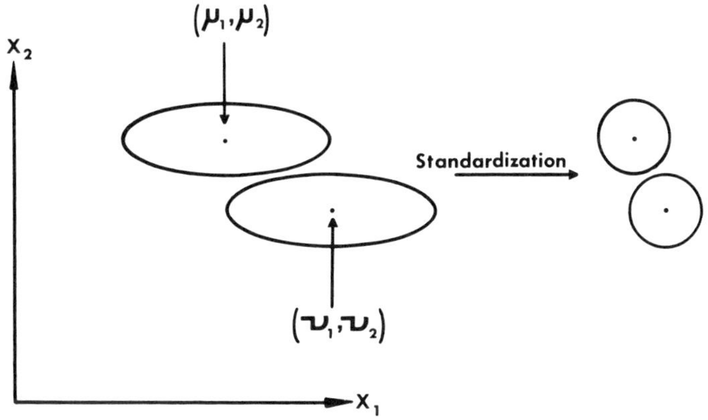

Figure 4.4 Effect of standardization on distance between class means

This is the classical approach to discriminant analysis. Its long history, beginning with Fisher (1936), has led to its being the most widely implemented technique in computer discriminant analysis packages. It should also be noted, as we demonstrate in Section 4.8, that for two normal classes with equal variance–covariance matrices the Fisher solution is asymptotically identical to the Bayes optimal solution (not so surprising when one recognizes that both work in terms of the first and second order moments).

In a study to discriminate between medical schools and other institutional users of the University of London Computer Centre on the basis of their computer usage several variables were used, amongst them the number of computational units used between 14 April and 11 May 1980 under two different operating systems. As in Chapter 2 we discarded institutions using no computer time under either of the operating systems and have taken natural logarithm transforms of the variables. The data are shown in Figure 4.5. The classical discriminant analysis decision surface is line *A*. The classification rule is

$$-0.297 \ln (\text{type 1 units}) + 0.640 \ln (\text{type 2 units}) - 2.478 \gtrless \ln \frac{21}{28} \Rightarrow \mathbf{x} \in \begin{cases} \Omega_1 \\ \Omega_2 \end{cases}$$

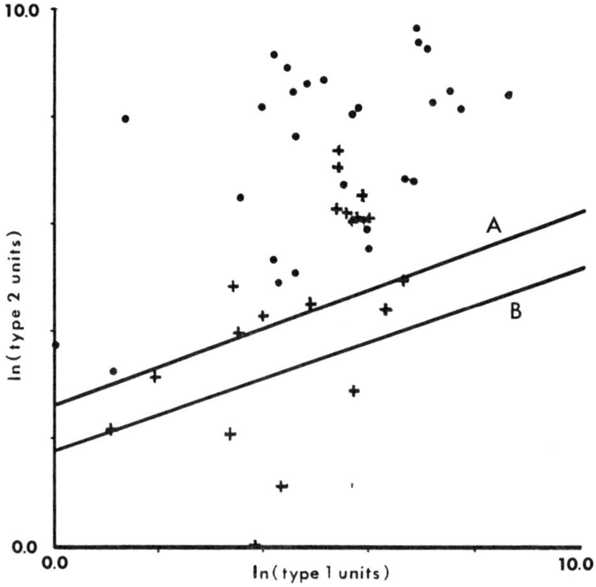

Figure 4.5 The computer usage data of Table 2.1. Institutions scoring 0 on either variable have been omitted. (a) Fisher's linear discriminant function. (b) The iterative steepest descent method

4.6 THE LEAST SQUARES APPROACH

We have tried to find a solution vector $\mathbf{w}$ such that as many as possible of the augmented sample vectors $\mathbf{y}_i$ satisfy

$$\mathbf{w}'\mathbf{y}_i > 0$$

If instead we try to find a vector $\mathbf{w}$ which satisfies

$$\mathbf{w}'\mathbf{y}_i = b_i$$

where the b_i are positive constants, one for each of the n sample points, then instead of needing the solution to a set of linear inequalities we must solve a set of linear equations. We can express the n equations in matrix form as

$$\underset{n \times (d+1)}{\mathbf{Y}} \quad \underset{(d+1) \times 1}{\mathbf{w}} = \underset{(n \times 1)}{\mathbf{B}}$$

where $\mathbf{Y} = (\mathbf{y}_1, \mathbf{y}_2, \ldots, \mathbf{y}_n)'$ and $\mathbf{B} = (b_1, \ldots, b_n)'$. For a given vector $\mathbf{B}$ we can apply standard least squares approaches which minimize the criterion

$$C_4 = (\mathbf{Yw} - \mathbf{B})'(\mathbf{Yw} - \mathbf{B})$$

to give a solution

$$\hat{\mathbf{w}} = (\mathbf{Y}'\mathbf{Y})^{-1}\mathbf{Y}'\mathbf{B} \tag{1}$$

There remains the question of choosing $\mathbf{B}$. One particularly interesting choice is

$$\mathbf{B} = \begin{bmatrix} \dfrac{n}{n_1} & \mathbf{u}_1 \\ \dfrac{n}{n_2} & \mathbf{u}_2 \end{bmatrix}$$

where $\mathbf{u}_i$ is a vector of n_i 1's and n_i is the number of design set vectors from class ω_i (so $n_1 + n_2 = n$).

To see why this is interesting let us write it out in full. We have

$$\mathbf{Y} = \begin{bmatrix} \mathbf{u}_1 & \mathbf{X}_1 \\ -\mathbf{u}_2 & -\mathbf{X}_2 \end{bmatrix}$$

with $\mathbf{X}_i'$ having as columns the class ω_i design set elements. Then (1), re-written as

$$(\mathbf{Y}'\mathbf{Y})\mathbf{w} = \mathbf{Y}'\mathbf{B}$$

becomes

$$\begin{bmatrix} n & n_1\bar{\mathbf{x}}_1' + n_2\bar{\mathbf{x}}_2' \\ n_1\bar{\mathbf{x}}_1 + n_2\bar{\mathbf{x}}_2 & \mathbf{X}_1'\mathbf{X}_1 + \mathbf{X}_2'\mathbf{X}_2 \end{bmatrix} \begin{bmatrix} v_0 \\ \mathbf{v} \end{bmatrix} = \begin{bmatrix} 0 \\ n(\bar{\mathbf{x}}_1 - \bar{\mathbf{x}}_2) \end{bmatrix}$$

From the top row of this matrix equation we obtain

$$v_0 = \left(-\frac{n_1}{n}\bar{\mathbf{x}}_1' - \frac{n_2}{n}\bar{\mathbf{x}}_2'\right)\mathbf{v} \tag{2}$$

The second row gives

$$(n_1\bar{\mathbf{x}}_1 + n_2\bar{\mathbf{x}}_2)v_0 + (\mathbf{X}_1'\mathbf{X}_1 + \mathbf{X}_2'\mathbf{X}_2)\mathbf{v} = n(\bar{\mathbf{x}}_1 - \bar{\mathbf{x}}_2) \tag{3}$$

Now, by using v_0 from (2) and the identity

$$n\mathbf{S} = \mathbf{X}_1'\mathbf{X}_1 + \mathbf{X}_2'\mathbf{X}_2 - n\bar{\mathbf{x}}_1\bar{\mathbf{x}}_1' - n_2\bar{\mathbf{x}}_2\bar{\mathbf{x}}_2'$$

where $\mathbf{S}$ is the estimate of assumed common variance–covariance matrix we have, from (3)

$$\frac{n_1 n_2}{n}(\bar{\mathbf{x}}_1 - \bar{\mathbf{x}}_2)(\bar{\mathbf{x}}_1 - \bar{\mathbf{x}}_2)'\mathbf{v} + n\mathbf{S}\mathbf{v} = n(\bar{\mathbf{x}}_1 - \bar{\mathbf{x}}_2)$$

Since, in the first term, $(\bar{\mathbf{x}}_1 - \bar{\mathbf{x}}_2)'\mathbf{v}$ is a scalar we have

$$n\mathbf{S}\mathbf{v} = (\text{scalar})(\bar{\mathbf{x}}_1 - \bar{\mathbf{x}}_2)$$

That is

$$\mathbf{v} \propto \mathbf{S}^{-1}(\bar{\mathbf{x}}_1 - \bar{\mathbf{x}}_2)$$

which is exactly the same solution we obtained in the last section for Fisher's criterion.

Another interesting choice for $\mathbf{B}$ is $\mathbf{B} = \mathbf{u}_n$, a vector of n 1's. Minimizing the sum of squares between $\mathbf{Yw}$ and this $\mathbf{B}$ is asymptotically equivalent to minimizing the mean square error between $\mathbf{vx} + v_0$ and the Bayes discriminant function

$$g(\mathbf{x}) = P(\omega_1|\mathbf{x}) - P(\omega_2|\mathbf{x})$$

One advantage of these least squares approaches is that they can easily be implemented on standard computer regression packages. Let us take the first case above as an example. We have

$$\begin{cases} \mathbf{w}'\mathbf{y}_i = n/n_1, & \text{for } \mathbf{y}_i \in \omega_1 \\ \mathbf{w}'\mathbf{y}_i = n/n_2, & \text{for } \mathbf{y}_i \in \omega_2 \end{cases}$$

That is

$$\begin{cases} \mathbf{v}'\mathbf{x}_i + v_0 = n/n_1, & \text{for } \mathbf{x}_i \in \omega_1 \\ -\mathbf{v}'\mathbf{x}_i - v_0 = n/n_2, & \text{for } \mathbf{x}_i \in \omega_2 \end{cases}$$

which in turn is

$$\begin{cases} \mathbf{v}'\mathbf{x}_i + v_0 = n/n_1, & \text{for } \mathbf{x}_i \in \omega_1 \\ \mathbf{v}'\mathbf{x}_i + v_0 = -n/n_2, & \text{for } \mathbf{x}_i \in \omega_2 \end{cases}$$

Thus, to implement this on a regression program one simply defines a criterion variable z by

$$z_i = \begin{cases} n/n_1, & \text{for } \mathbf{x}_i \in \omega_1 \\ -n/n_2, & \text{for } \mathbf{x}_i \in \omega_2 \end{cases}$$

and then regresses z on $(x_1, \ldots, x_d)$. Applying this to the computer usage data of the preceding section, with

$$z_i = \begin{cases} 60/35 = 1.71, & \text{for } \mathbf{x}_i \in \omega_1 \\ -60/25 = -2.40, & \text{for } \mathbf{x}_i \in \omega_2 \end{cases}$$

gives the same discriminant function as Fisher's method.

One of the differences in emphasis between pattern recognition and statistical work on discrimination lies in the notion of linear separability and how important it is that the final weight vector should separate the samples perfectly when this is possible. In statistical work little importance has been attached to this but in pattern recognition work, especially during the early stages, considerable stress was laid on it. This is perhaps hinted at by the proof in Section 4.4 that if the samples are linearly separable then the error correction method outlined is guaranteed to find a solution vector which separates the samples but that if the data are non-separable then unfortunately the constraints on the choice of ρ_k prevent the simple error correction method from converging. We now illustrate this emphasis on separability further by presenting a method which both guarantees that the solution will separate the samples when this is possible and always converges to a solution. First observe that, in contrast to the error correction method, the least squares methods always yield a final solution—but with the choices of $\mathbf{B}$ we have so far discussed there is no guarantee of separation whenever this is possible.

There is, however, no reason why we should not try to minimize C_4 by varying both $\mathbf{w}$ and $\mathbf{B}$, subject to the constraint that $\mathbf{B} > \mathbf{0}$ of course. Several approaches along these lines have been suggested (see, for example, Duda and Hart, 1973, and Ho and Kashyap, 1965, 1966) many of them based on an iterative steepest descent method using

$$\nabla_{\mathbf{w}} C_4 = 2\mathbf{Y}'(\mathbf{Yw} - \mathbf{B})$$

and

$$\nabla_{\mathbf{B}} C_4 = -2(\mathbf{Yw} - \mathbf{B})$$

The method given here also has its origin in these derivatives. First, note that if we begin with an arbitrary $\mathbf{B} = \mathbf{B}_1 > \mathbf{0}$ and only *increase* components of $\mathbf{B}$ then $\mathbf{B}_k$, the estimate of $\mathbf{B}$ at the kth iterative step, will remain positive. Moreover, if we only increase the components of $\mathbf{B}$ for which C_4 has a negative gradient, then C_4 will decrease. We now prove the following theorem.

Theorem

For the iterative algorithm given by

$$\mathbf{B}_1 > \mathbf{0}, \quad \text{but otherwise arbitrary}$$
$$\mathbf{w}_1, \quad \text{arbitrary}$$
$$\mathbf{B}_{k+1} = \mathbf{B}_k + (\mathbf{Y}\mathbf{w}_k - \mathbf{B}_k) + |\mathbf{Y}\mathbf{w}_k - \mathbf{B}_k|$$

and

$$\mathbf{w}_{k+1} = \mathbf{w}_k + \rho\mathbf{Y}'|\mathbf{Y}\mathbf{w}_k - \mathbf{B}_k|$$

where ρ satisfies $0 < \rho < 2/\lambda$ with λ the maximum eigenvalue of $\mathbf{Y}'\mathbf{Y}$:

(a) if the two design samples are linearly separable then the algorithm will converge in a finite number of steps, and

(b) if either $\mathbf{Y}'|\mathbf{Y}\mathbf{w}_k - \mathbf{B}_k| = 0$ with $(\mathbf{Y}\mathbf{w}_k - \mathbf{B}_k) \neq 0$ or $\mathbf{Y}'|\mathbf{Y}\mathbf{w}_k - \mathbf{B}_k|$ converges to zero while $(\mathbf{Y}\mathbf{w}_k - \mathbf{B}_k)$ converges to some non-zero value, then the samples are not linearly separable.

Proof. To simplify the notation define

$$\mathbf{e}_k = \mathbf{Y}\mathbf{w}_k - \mathbf{B}_k$$

so that

$$\begin{aligned}\mathbf{e}_{k+1} &= \mathbf{Y}\mathbf{w}_{k+1} - \mathbf{B}_{k+1} \\ &= \mathbf{Y}\mathbf{w}_k + \rho\mathbf{Y}\mathbf{Y}'|\mathbf{e}_k| - \mathbf{B}_k - \mathbf{e}_k - |\mathbf{e}_k| \\ &= (\rho\mathbf{Y}\mathbf{Y}' - \mathbf{I})|\mathbf{e}_k|\end{aligned}$$

From this

$$\mathbf{e}'_{k+1}\mathbf{e}_{k+1} = |\mathbf{e}_k|'(\rho^2\mathbf{Y}\mathbf{Y}'\mathbf{Y}\mathbf{Y}' - 2\rho\mathbf{Y}\mathbf{Y}' + \mathbf{I})|\mathbf{e}_k|$$

which implies

$$\mathbf{e}'_{k+1}\mathbf{e}_{k+1} = \mathbf{e}'_k\mathbf{e}_k - (\mathbf{Y}'|\mathbf{e}_k|)'(2\rho\mathbf{I} - \rho^2\mathbf{Y}'\mathbf{Y})(\mathbf{Y}'|\mathbf{e}_k|)$$

Now if we choose ρ such that $0 < \rho < 2/\lambda$ the matrix $(2\rho\mathbf{I} - \rho^2\mathbf{Y}'\mathbf{Y})$ will be positive definite. This means that when

$$\mathbf{Y}'|\mathbf{e}_k| \neq 0$$

then

$$\mathbf{e}'_{k+1}\mathbf{e}_{k+1} < \mathbf{e}'_k\mathbf{e}_k$$

That is, $\mathbf{e}'_k\mathbf{e}_k$ is monotonically decreasing. (This simply says that, provided we do not step too far, any step in the direction of the negative gradient must decrease the criterion function.) Moreover, since $\mathbf{e}'_k\mathbf{e}_k \geq 0$ this sequence must converge.

So far we have not distinguished between the separable and non-separable cases. Let us first take (a) in the statement of the theorem and let $\hat{\mathbf{w}}$ and $\hat{\mathbf{B}}$ be solutions.

Since $\mathbf{e}_k'\mathbf{e}_k$ converges, the correction term $\mathbf{Y}'|\mathbf{e}_k|$ must converge to zero. Now if $|\mathbf{e}_k| \neq 0$ then

$$|\mathbf{e}_k|'\mathbf{Y}\hat{\mathbf{w}} = |\mathbf{e}_k|'\hat{\mathbf{B}} > 0$$

(since $\hat{\mathbf{B}} > \mathbf{0}$) so that $|\mathbf{e}_k|'\mathbf{Y} \neq 0$. Thus, for $\mathbf{Y}'|\mathbf{e}_k|$ to converge to zero we must have $|\mathbf{e}_k| \to \mathbf{0}$ and hence $\mathbf{e}_k \to \mathbf{0}$. However, since $\mathbf{B}_k$ starts positive and its components never decrease the only way for $\mathbf{e}_k$ to converge to zero is if $\mathbf{w}_k$ converges to a vector $\hat{\mathbf{w}}$ satisfying

$$\mathbf{Y}\hat{\mathbf{w}} = \hat{\mathbf{B}}$$

Furthermore, after some finite number of steps we must have

$$\sqrt{\mathbf{e}_k'\mathbf{e}_k} < \min_i B_{ki}$$

(where B_{ki} is the ith component of $\mathbf{B}_k$) so that

$$\mathbf{Y}\mathbf{w}_k = \mathbf{B}_k + \mathbf{e}_k > \mathbf{0}$$

(by positivity of $\mathbf{B}_k$) which is thus a solution.

In the separable case we know that $\mathbf{Y}'|\mathbf{e}_k| = \mathbf{0}$ only if $\mathbf{e}_k = \mathbf{0}$ and in the non-separable case we know that $\mathbf{e}_k \neq \mathbf{0}$ and does not converge to zero. Thus, if we find $\mathbf{Y}'|\mathbf{e}_k| = \mathbf{0}$ with $\mathbf{e}_k \neq \mathbf{0}$ or $\mathbf{Y}'|\mathbf{e}_k| \to \mathbf{0}$ with $\mathbf{e}_k \to \boldsymbol{\delta} \neq \mathbf{0}$ we have evidence of non-separability. ■

Whether the conditions for identifying non-separability are useful in practice will depend on the problem.

To illustrate the algorithm it was first applied to the separable data of Table 4.1 (Figure 4.1) which is explained in Section 4.3. The $\mathbf{Y}'\mathbf{Y}$ matrix for this data is

$$\mathbf{Y}'\mathbf{Y} = \begin{bmatrix} 247 & 218 & 76 \\ 218 & 232 & 72 \\ 76 & 72 & 26 \end{bmatrix}$$

with largest eigenvalue $\lambda = 481$. Using

$$\rho = 2.0/(\lambda + 0.0001) < 2.0/\lambda$$

and initial $\mathbf{w}$ and $\mathbf{B}$ with every component unity led to solution vector

$$\mathbf{w}' = (-14.3, 21.9, -19.7)$$

with $\mathbf{B}' = (35.5, 27.8, 23.5, 21.7, 23.5, 22.1, 19.3, 18.6, 18.0, 15.2, 8.4, 1.0, 1.0, 5.9, 10.9, 13.3, 12.8, 16.8, 18.8, 16.8, 24.7, 23.8, 45.9, 47.0, 49.0, 64.7)$. As expected, all the B_i are positive.

To illustrate the algorithm on non-separable data it was applied to the computer usage data of Table 2.1. This data set is also described in Section 4.5. Computer time requirements for the algorithm on this data set proved excessive and it was necessary to terminate the process before either of the non-separability conditions had become apparent. The decision surface estimate when the process was stopped is line *B* in Figure 4.5. Its equation is

$$0 = -3.66 \ln (\text{type 1 units}) + 8.23 \ln (\text{type 2 units}) - 18.94 - \ln \frac{P(\omega_2)}{P(\omega_1)}$$

4.7 CHOOSING AN ESTIMATION METHOD

We have presented several methods for estimating parameters of a linear discriminant function. Unfortunately, none of these methods is an overall best—they have advantages in some ways and disadvantages in others. For example, linear programming methods do not have the problems of finding a suitable step size or of choosing a stopping rule. Both of these problems are typical of steepest descent and error correction methods. On the other hand accumulated computational error can be a problem when using linear programming while adaptive methods can be viewed as feedback processes, exponentially damping any computational errors. The linear programming method outlined here guarantees convergence in finite time to a solution minimizing the perceptron criterion whether the data are linearly separable or not, but, as we have already commented, the linear programming formulation of a problem might be very large, especially if n is big. If it is important that a separating decision surface should be found when the samples are linearly separable then an error correction or linear programming or the least squares method where $\mathbf{B}$ is also found (as outlined at the end of Section 4.6) should be used. However, one should also bear in mind that the error correction method does not converge if the samples are non-separable. Fisher's linear discriminant method has the advantage that use can easily be made of incomplete design set elements (they do not have to be discarded). Thus, any elements for which x_i and x_j are present can contribute towards the estimate of the correlation between the ith and jth variables, regardless of whether or not other components are present. A similar remark applies to the estimates of the means $\bar{\mathbf{x}}_1$ and $\bar{\mathbf{x}}_2$. (But see Chapter 8.)

Standard computer packages can be used for the regression, classical discriminant analysis, and linear programming methods, but the error correction type of approaches are very easy to program in any scientific programming language.

In choosing between methods one should also be cognizant of the different criteria optimized—that is, whether a particular method uses the least squares criterion, or whatever.

Although no single method provides an overall best, most of the methods can yield good results if carefully used. The most important recommendation is that the user should be aware of the weaknesses of the chosen method so that caution can be exercised when appropriate.

4.8 SPECIALIZATIONS AND GENERALIZATIONS

We have already commented on the importance of the special case of normal class-conditional pdfs. If, as suggested in Section 4.1, we use as discriminant function

$$g(\mathbf{x}) = \ln\{p(\mathbf{x}|\omega_1)/p(\mathbf{x}|\omega_2)\}$$

then we have

$$g(\mathbf{x}) = -\tfrac{1}{2}(\mathbf{x} - \boldsymbol{\mu}_1)'\boldsymbol{\Sigma}_1^{-1}(\mathbf{x} - \boldsymbol{\mu}_1) + \tfrac{1}{2}(\mathbf{x} - \boldsymbol{\mu}_2)'\boldsymbol{\Sigma}_2^{-1}(\mathbf{x} - \boldsymbol{\mu}_2)$$
$$+ \tfrac{1}{2}\ln|\boldsymbol{\Sigma}_2|/|\boldsymbol{\Sigma}_1| \gtrless \ln k \Rightarrow \mathbf{x} \in \begin{cases}\Omega_1\\ \Omega_2\end{cases}$$

This is a quadratic function of the components of $\mathbf{x}$ and for the special case $\boldsymbol{\Sigma}_1 = \boldsymbol{\Sigma}_2 = \boldsymbol{\Sigma}$ reduces to

$$\mathbf{x}'\boldsymbol{\Sigma}^{-1}(\boldsymbol{\mu}_1 - \boldsymbol{\mu}_2) - \tfrac{1}{2}\boldsymbol{\mu}_1'\boldsymbol{\Sigma}^{-1}\boldsymbol{\mu}_1 + \tfrac{1}{2}\boldsymbol{\mu}_2'\boldsymbol{\Sigma}^{-1}\boldsymbol{\mu}_2 \gtrless \ln \hat{k} \Rightarrow \mathbf{x} \in \begin{cases}\Omega_1\\ \Omega_2\end{cases}$$

Thus, for normal classes with identical variance–covariance matrices the optimal decision rule is linear.

Note also that this gives, once again,

$$\mathbf{v} = \boldsymbol{\Sigma}^{-1}(\boldsymbol{\mu}_1 - \boldsymbol{\mu}_2)$$

in the notation of earlier sections.

So far the discussion has been in terms of two classes. Several ways have been proposed to extend the methods to the general N class case, for example:

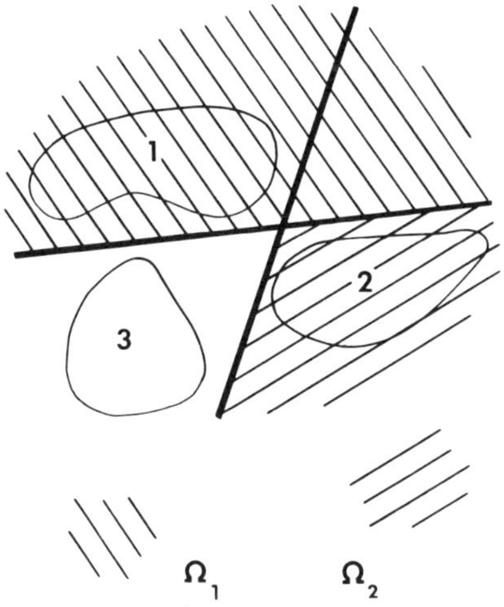

Figure 4.6 N–1 two-class decision surfaces, each separating Ω_i from $(\Omega_{i+1}, \ldots, \Omega_N)$ $(i = 1, \ldots, N - 1)$

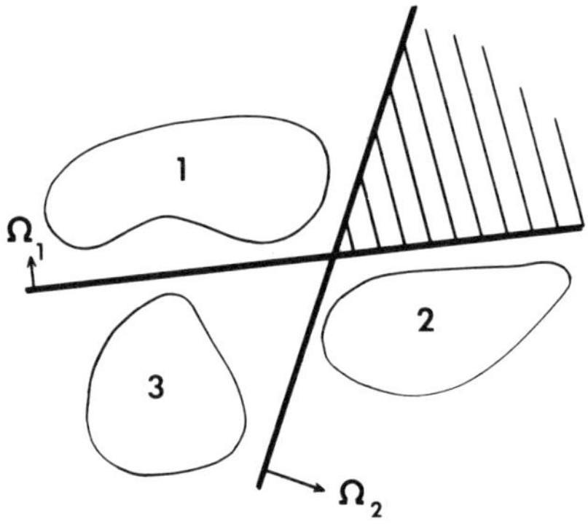

Figure 4.7 $N-1$ two-class decision surfaces, each separating Ω_i from $(\Omega_1, \ldots, \Omega_{i-1}, \Omega_{i+1}, \ldots, \Omega_n)$ $(i = 1, \ldots, N-1)$

(i) We could use $(N - 1)$ two-class decision surfaces, the first separating Ω_1 from $\Omega_2, \ldots, \Omega_N$; the second separating Ω_2 from $\Omega_3, \ldots, \Omega_N$; etc. A disadvantage of this is that some ordering has to be imposed on the classes and the lower the index the greater the proportion of the space assigned to that class (Figure 4.6).

(ii) We could use $(N - 1)$ two-class decision rules, each one separating Ω_i $(i = 1, \ldots, N - 1)$ from all Ω_j $(j = 1, \ldots, N; j \neq i)$. However, this rule can lead to undecided regions—the shaded area in Figure 4.7 is both Ω_1 and Ω_2.

(iii) We could use $N(N - 1)/2$ two-class decision rules, one for each pair of classes. Apart from possible excessive computational requirements this also has the disadvantage that it can lead to undecided regions (the shaded area in Figure 4.8).

(iv) The most widely adopted method is a more fundamental general-

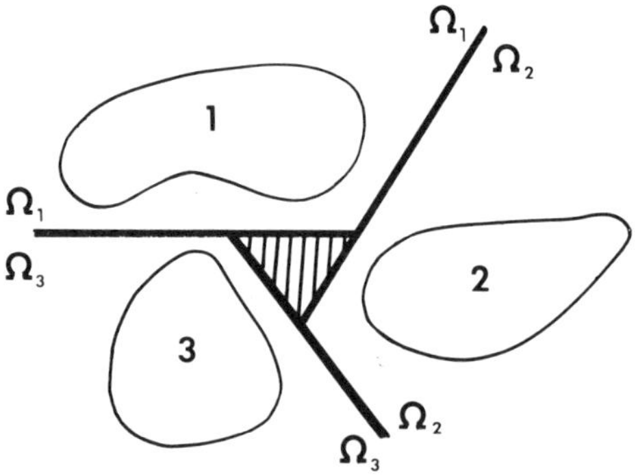

Figure 4.8 $N(N - 1)/2$ decision surfaces, one for each pair of classes

ization of the two-class case. In Section 4.1 we put

$$\frac{p(\omega_1|\mathbf{x})}{p(\omega_2|\mathbf{x})} \gtrless 1 \Rightarrow \mathbf{x} \in \begin{cases} \Omega_1 \\ \Omega_2 \end{cases}$$

where for convenience we have taken $k = 1$ (the extension to general k is straightforward). This can be rewritten as

$$p(\omega_1|\mathbf{x}) - p(\omega_2|\mathbf{x}) \gtrless 0 \Rightarrow \mathbf{x} \in \begin{cases} \Omega_1 \\ \Omega_2 \end{cases}$$

This, in turn, can be immediately generalized as

$$p(\omega_i|\mathbf{x}) = \max_j p(\omega_j|\mathbf{x}) \Rightarrow \mathbf{x} \in \Omega_i$$

(See also Section 1.3.) Once again any set of functions $q_i(\mathbf{x})$ $(i = 1, \ldots, N)$ for which

$$q_i(\mathbf{x}) = \max_j q_j(\mathbf{x}) \Leftrightarrow p(\omega_i|\mathbf{x}) = \max_j p(\omega_j|\mathbf{x}) \tag{1}$$

will give the same classification regions Ω_i. We could exploit the freedom of choice in $q_i(\mathbf{x})$ by attempting to find linear functions satisfying (1). More generally, however (and compare Section 4.1), we could assume linearity of the differences $q_i(\mathbf{x}) - q_j(\mathbf{x})$. (Linearity of the $q_i(\mathbf{x})$ implies linearity of the differences whereas the converse does not hold.) This leads to $N(N-1)/2$ linear functions

$$q_i(\mathbf{x}) - q_j(\mathbf{x}) = \sum \alpha_k^{(ij)} x_k$$

Fortunately we can exploit the relationships between the $q_i(\mathbf{x}) - q_j(\mathbf{x})$ to reduce the number to $(N-1)$ linear functions. Thus

$$(q_i - q_j) = (q_i - q_N) - (q_j - q_N)$$

We therefore require

$$q_i(\mathbf{x}) - q_N(\mathbf{x}) = \sum \alpha_k^{(i)} x_k \quad (i = 1, \ldots, N-1)$$

leading to the classification rule:

If $\sum \alpha_k^{(i)} x_k < 0$ for all $i = 1, \ldots, N-1$, then $\mathbf{x} \in \Omega_N$.
Otherwise, $\sum \alpha_k^{(i)} x_k = \max_j \sum \alpha_k^{(j)} x_k \Rightarrow \mathbf{x} \in \Omega_i$.

Common q functions are $q_i(\mathbf{x}) = \ln p(\omega_i|\mathbf{x})$ (again, as in Section 4.1, by the monotonicity of the ln function, (1) is true). These give

$$q_i(\mathbf{x}) - q_N(\mathbf{x}) = \sum \alpha_k^{(i)} x_k$$

hence

$$\ln p(\omega_i|\mathbf{x}) - \ln p(\omega_N|\mathbf{x}) = \sum \alpha_k^{(i)} x_k$$

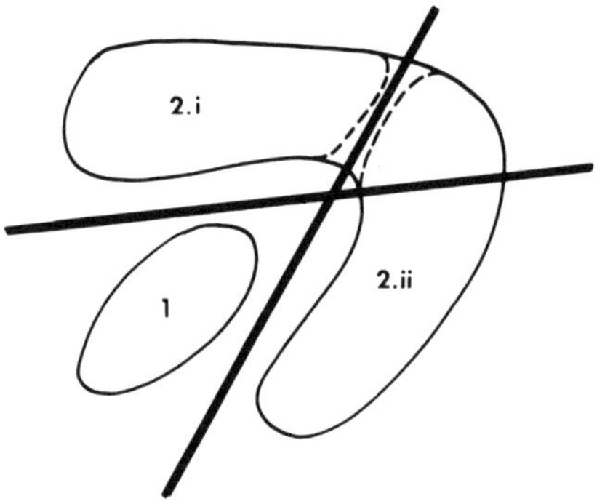

Figure 4.9 Splitting non-convex classes into subclasses

so that

$$p(\omega_i|\mathbf{x}) = p(\omega_N|\mathbf{x})\exp(\Sigma\, \alpha_k^{(i)} x_k)$$

and

$$p(\omega_N|\mathbf{x}) = \left[1 + \sum_{i=1}^{N-1} \exp(\Sigma\, \alpha_k^{(i)} x_k)\right]^{-1}$$

For the particular case of Fisher's linear discriminant function (Section 4.5), Section 6.5 presents the usual multicategory extension.

The decision surface resulting from these rules is *piecewise linear*. The regions Ω_i generated by linear discriminant functions are easily seen to be convex. Unfortunately, however, the Bayes optimal regions are not always restricted in this way. One approach to tackling the problem of non-convex classes is to split them into subclasses (Figure 4.9), but this is difficult in high dimensional spaces. Another approach is to use generalized linear discriminant functions, as discussed in Section 4.1.

4.9 CONCLUSIONS

For several reasons, linear discriminant functions have been the focus of a considerable amount of research effort. First there is their analytic simplicity. Secondly, they represent the simplest special case (of more general discriminant functions) which has wide applicability. Furthermore, more complicated functions may not be feasible if the dimensionality is large.

The statistical work in this area stems from Fisher's paper of (1936), using the method outlined in Section 4.5. This method and its developments are the primary concern of Lachenbruch (1975), who provides an introduction to the method and its properties as well as giving a summary of more recent advances. The pattern recognition work has its origin in neuronal network models—for example, McCulloch and Pitts (1943), and Rosenblatt (1962). An important early book on linear discriminant functions in pattern recognition is that of Nilsson (1965) and an excellent survey may be found in Duda and Hart (1973, ch. 5).

EXERCISES

4.1 Write a general computer program to minimize the perceptron criterion using an error correction algorithm.

4.2 Generate random samples from two multivariate normal classes with unequal variance–covariance matrices and apply the error correction program written in 4.1 using a generalized linear decision surface which includes second-order moments.

4.3 Using a standard linear programming computer package and a standard regression analysis package compare the computer time requirements for several different data sets.

4.4 In Chapter 5 we show that if the class-conditional pdfs are defined by independent binary variables then a linear decision surface is optimal. Investigate the use of error correction and linear programming algorithms with binary variables. Might other kinds of mathematical programming be applied?

4.5 If there are $N > 2$ classes then using method (iv) of Section 4.8 we can define a decision rule by

$$\mathbf{w}_i'\mathbf{x} > \mathbf{w}_j'\mathbf{x} \quad \text{for all } j \neq i \Rightarrow \mathbf{x} \in \Omega_i$$

where $\mathbf{w}_i$ is the weight vector for class ω_i. To estimate the $\mathbf{w}_i$ an error correction rule analogous to that of Section 4.4 is as follows.

Present the design set points sequentially to the classifier and let $\mathbf{w}_{ik}$ be the estimate of $\mathbf{w}_i$ at the kth presentation. Suppose that a point $\mathbf{u}$, from class ω_i is next presented. Then if

$$\mathbf{w}_{ik}'\mathbf{u} > \mathbf{w}_{jk}'\mathbf{u}, \quad \text{for all } j \neq i$$

(i.e. if $\mathbf{u}$ is correctly classified) let

$$\mathbf{w}_{ik+1} = \mathbf{w}_{ik}, \quad \text{for all } i.$$

However, if

$$\mathbf{w}_{qk}'\mathbf{u} > \mathbf{w}_{ik}'\mathbf{u}, \quad \text{for } q = l, m, \ldots, p$$

(i.e. if $\mathbf{u}$ is incorrectly classified) let

$$\begin{aligned}
\mathbf{w}_{ik+1} &= \mathbf{w}_{ik} + \rho\mathbf{u}, \\
\mathbf{w}_{qk+1} &= \mathbf{w}_{qk} - \rho\mathbf{u}, \quad q = l, m, \ldots, p \\
\mathbf{w}_{jk+1} &= \mathbf{w}_{jk}, \qquad \text{for all } j \neq i, l, m, \ldots, p
\end{aligned}$$

Show that if the classes are linearly separable this rule will find a solution.

CHAPTER 5

Discrete Variables

5.1 INTRODUCTION

The major part of this book is concerned with continuous variables, i.e. with variables which can take any value from an interval of real numbers. Sometimes, however, variables are used which can only take a finite number of values. It is with such variables that this chapter is concerned.

To take an example, in an attempt to develop a screening questionnaire to identify people who are at risk from dental caries one might ask questions about dietary preferences. Thus, one might ask if the respondents liked ice-cream a lot, a little, or not at all. Instead of an infinite number of possible responses we have only three. Moreover, there is an additional complication: how should we assess the relative merit of the responses? Should, for example, 'a lot' be coded as 2, 'a little' as 1, and 'not at all' as 0? Or perhaps the coding 8, 0, -1 would be better. Clearly this example has introduced problems beyond those normally encountered when dealing with continuous variables (though this is not always true. Sometimes there is a case for a non-linear scaling (say, by a log transform) of a continuous variable before using it). In some situations these problems disappear if a sufficiently flexible decision surface (see Chapter 1) is used but in many others they do not. It is vital for the investigator to think carefully about the nature of the variables and the meanings of their scales. This point is emphasized by an additional problem which can arise with categorical variables. For the attempted vulnerability to dental caries classification another variable might be race, with possible 'values' Caucasoid, Negroid, Mongoloid, or Other. Now, unlike the ice-cream variable above, it is evident that this variable does not even have an intrinsic ordering. Clearly, whatever difficulties are encountered in discriminations and classifications involving variables of the first type ('ordinal' variables), even worse difficulties will be encountered with the second type ('nominal' variables).

The astute reader may object that even putative continuous variables are really discrete. After all, one may measure someone's height to the nearest centimetre or millimetre, but there is little point in making finer divisions.

This is, of course, true and one could indeed treat continuous variables as discrete variables with a very large number of possible categories. And it is in this last qualification that the practical difference between the two situations lies. If there are a very large number of divisions then not only would practical difficulties be encountered in applying the methods of categorical variables but also there is no effective loss of accuracy in making the continuity approximation. This is a result of the small size of each of the categories coupled with the problems of measurement error and certain implicit assumptions about the correlations of neighbouring cells.

One can also sometimes take the opposite view: if the categories are ordered then one can treat them as points on a continuous scale. Problems such as that illustrated by the ice-cream example, about which points to use, will be encountered and this says nothing about the case of unordered categories. (Despite these difficulties it is not uncommon to find discrete variables being treated as continuous and vice versa. This is returned to briefly in Section 5.4.)

We can imagine a continuum of categorical variables with those with an infinite number of ordered categories (continuous variables) at one extreme. At the opposite extreme lies the binary or dichotomous variable, which has only two categories. Binary variables are very important. The avoid the problems of different inter-category intervals illustrated in the ice-cream example—suitable rescaling reduces them to values 0 and 1. Similarly, difficulties due to arbitrarily imposed orders simply do not arise. Moreover, they can be used to resolve the problems of non-ordered categories in variables which can take more than two values by replacing the g categories by g binary variables. Thus, in the example above the categories would be replaced by four variables

v_1 scoring 1 if Caucasoid; 0 otherwise.
v_2 scoring 1 if Negroid; 0 otherwise.
v_3 scoring 1 if Mongoloid; 0 otherwise.
v_4 scoring 1 if Other, 0 otherwise.

It is easy to see that by this process linear decision surfaces can be defined which separate any specified subset of categories from the remainder. Thus, in the general case of a variable with g categories we can define

$$l = a_1 v_1 + \ldots + a_g v_g$$

where $v_i = 1$ if the object falls in the ith category, and 0 otherwise, and

$$a_i = \begin{cases} +1, & \text{for those categories in the specified subset} \\ -1, & \text{otherwise} \end{cases}$$

Then $l = +1$ for an object falling in the specified subset and $l = -1$ otherwise. l can then be combined with any other variables in the study.

There are other ways in which categorical variables may be recoded and treated as several binary variables, but in general caution should be exercised.

Any structure implicit in the multivariate representation must not conflict with the discrimination method adopted. (Thus, for example, the fact that the v_i above are related—a 1 in one means a 0 in the others—could sometimes cause difficulties.)

A second reason for the importance of binary variables lies in the frequent splitting of classes into two groups: those possessing some attribute, and all the others. This simplest of classifications has also had an impact through the binary basis of electronic computers.

Because of the importance and fundamental nature of binary variables much of the subject matter of this chapter is expressed in terms of such variables. Ways of extending to variables with more than two categories are usually straightforward and such extensions are demonstrated.

As has already been hinted at by the choice of examples, discrimination and classification problems involving categorical variables arise often in the behavioural sciences (psychology, sociology, etc.) where the measurements of the physical sciences are matched by questions and questionnaires.

A problem which has been referred to repeatedly in other contexts in this book is that of the curse of dimensionality. Here it simply means that the number of cells increases exponentially with the number of variables measured. This has two chief consequences: first, sparse distributions of observations, with many of the cells empty, are common even though a large number of observations may be available: secondly, if there are G cells then $(G - 1)$ parameters are needed to define the distribution completely. These points are discussed in Sections 5.2 and 5.3.

5.2 DISTRIBUTION-FREE METHODS

As has already been explained in Chapter 1, since the optimal decision surfaces are described in terms of ratios of probabilities a reasonable way to design a classification rule is by comparing estimates of the probability functions of the two (or more) classes concerned (weighted appropriately by prior probabilities and the costs of the different types of misclassification). For continuous variables methods of pdf estimation were discussed in Chapters 2 and 3. In this section we shall discuss corresponding distribution-free methods for categorical variables and in Section 5.3 corresponding parametric methods. During these discussions it will become apparent that in some cases explicit pdf estimation can be avoided and we can proceed immediately to estimate the rank order of the various probabilities concerned (cf Section 2.4).

For categorical variables there is a possibility which is not present for continuous variables. Namely, that in principle we can estimate the class-conditional probability distribution functions without making any assumptions at all about their shape or form (even for non-parametric continuous variable methods a value for the spread parameter h must be chosen). The distribution of observations over the cross-classifications imposed by the categorical variables is, after all, simply an empirical realization of a multinomial distribution. We could therefore use the proportion of objects from class i falling in a cell as an estimate of the probability that class i objects have of falling in that cell. This idea forms the starting point for our discussion.

5.2.1 The multinomial solution

Let us suppose that there are d variables with the ith one having g_i possible categories and let

$$G = \prod_{i=1}^{d} g_i$$

Suppose also that measurements have been taken on n_j objects from class j ($j = 1, \ldots, N$) and that n_{kj} of these fall in cell k ($k = 1, \ldots, G$). Then an obvious estimate of the probability that an object from class j will fall in cell k is given by

$$\hat{P}_{kj} = n_{kj}/n_j$$

These estimates can then be substituted in the decision rules of Chapter 1.

Recall that we have in effect already considered this estimator in Chapter 2 where we suggested dividing each continuous variable into a number of discrete groups to form the histogram estimator. The present situation is exactly the same, except that we do not have to form groups—they are already defined by the nature of the variables. This identity implies that the multinomial approach has all the disadvantages of the histogram approach. To illustrate these suppose that each of the d variables has g categories so that $G = g^d$; that is, the number of cells increases exponentially with d. In a set of data to be used as an example below, there are 30 binary variables. For reliable estimation of the probabilities in each cell we might require an average of ten observations per cell. That is approximately 10^{10} observations. Since in this example each observation vector is a series of 30 questions asked of a person coupled with an interview by a psychiatrist, effective estimation of the decision rule requires interviews with more people than there are currently alive on the Earth. Not a very useful approach!

If fewer observations are taken then many of the cells will be empty and few will have sufficient observations to make the probability estimates reliable. A sparse distribution of observations means that most future classifications would have to be made by comparing a probability estimate of 0 for one class with an estimate of 0 for the other class. Not the most fruitful of decision rules! (The question of robustness of the multinomial method is also considered in Chapter 6.)

The conclusion is that only for small G will the multinomial approach be feasible.

5.2.2 Kernel methods

Aitchison and Aitken (1976) extended the continuous form of the kernel estimator (Section 2.3) to the categorical variable case. We shall follow their development, beginning with the special case of binary variables and showing how the method may be extended to nominal and ordinal variables with more than two categories.

As pointed out above, a disadvantage of the straightforward multinomial approach is that it can require an impossibly large number of observations. The kernel method 'smooths out' the observed sample so that the estimate of probability in any cell depends not only on observations which may fall in that cell, but also on observations which fall in nearby cells. (Thus, it is directly analogous to the continuous case, where the discrete probability spikes at the sample points are subjected to a convolution smoothing process.)

The general form for the kernel estimator is

$$\hat{p}(\mathbf{x}) = \frac{1}{n} \sum_{i=1}^{n} K(\mathbf{x} - \mathbf{x}_i)$$

where $\mathbf{x}$ is the point at which the estimate is desired and $\{\mathbf{x}_1, \ldots, \mathbf{x}_n\}$ is the design set for this class. $\hat{p}(\mathbf{x})$ is thus seen to be an average of the contributions from each $\mathbf{x}_i$, where each contribution is given by $K(\mathbf{x} - \mathbf{x}_i)$. Note that if

$$K(\mathbf{x} - \mathbf{x}_i) = \begin{cases} 1, & \text{if } \mathbf{x} = \mathbf{x}_i \\ 0, & \text{otherwise} \end{cases}$$

that is, with no smoothing, then we have the multinomial estimator.

Just as with the continuous case it is necessary to decide how much smoothing should be used and also, just as with the continuous case, it is convenient to split this question into two parts:

(a) What is the functional form relating distance from a sample point to the weight contributed by that point? (The form of K for the continuous case.)

(b) What value should be chosen for the spread parameter of (a)? (The value of h in the continuous case.)

Aitchison and Aitken used

$$K(\mathbf{x} - \mathbf{y}) = \prod_{i=1}^{d} K_0(x_i - y_i)$$

with

$$K_0(x_i - y_i) = \begin{cases} \lambda, & \text{if } x_i = y_i \\ (1 - \lambda), & \text{if } x_i \neq y_i \end{cases}$$

λ here (with $\frac{1}{2} \leq \lambda \leq 1$) serves in a role similar to, though not identical to, the spread parameter, h, of the continous case.

Further insight into this form of K can be obtained if we rewrite it as

$$K(\mathbf{x} - \mathbf{y}) = \prod_{i=1}^{d} \lambda^{1-\delta(x_i - y_i)}(1 - \lambda)^{\delta(x_i - y_i)}$$

(where $\delta(x_i - y_i) = 1$ if $x_i \neq y_i$; 0 if $x_i = y_i$). Then we have

$$K(\mathbf{x} - \mathbf{y}) = \lambda^{d-r(\mathbf{x}-\mathbf{y})}(1 - \lambda)^{r(\mathbf{x}-\mathbf{y})}$$

where $r(\mathbf{x} - \mathbf{y}) = (\mathbf{x} - \mathbf{y})'(\mathbf{x} - \mathbf{y})$, the number of components in which $\mathbf{x}$ and $\mathbf{y}$ disagree. A sample point at $\mathbf{x}$ is now seen to give weight λ^d at $\mathbf{x}$ itself, $\lambda^{d-1}(1 - \lambda)$ at those d cells which differ from $\mathbf{x}$ in only one component, $\lambda^{d-2}(1 - \lambda)^2$ at those $d(d - 1)/2$ cells which differ from $\mathbf{x}$ in two components, and so on.

From this one can see the necessity of the restrictions that $\lambda \geqslant \frac{1}{2}$ (so that design set points close to $\mathbf{x}$ contribute more to the probability estimate at $\mathbf{x}$ than those far away) and that $\lambda \leqslant 1$ (so that $\hat{p}(\mathbf{x})$ is not negative).

It only remains to choose the value of λ. Aitchison and Aitken recommend the jackknife method as yielding a good value. This involves maximizing

$$W(\lambda | X) = \prod_{i=1}^{n} \hat{p}(\mathbf{x}_i | X - \mathbf{x}_i, \lambda)$$

where $\hat{p}(\mathbf{x}_i | X - \mathbf{x}_i, \lambda)$ is the estimate of the probability at $\mathbf{x}_i$ based on the sample $X - \mathbf{x}_i = \{\mathbf{x}_i, \ldots, \mathbf{x}_{i-1}, \mathbf{x}_{i+1}, \ldots, \mathbf{x}_n\}$.

To extend this method to variables with more than two categories we have, for nominal variables

$$K_0(x_i - y_i) = \begin{cases} \lambda, & \text{if } x_i = y_i \\ \dfrac{1 - \lambda}{g_i - 1}, & \text{if } x_i \neq y_i \end{cases}$$

(where, as before, g_i is the number of categories of the ith variable).

For ordinal variables with, for example, $g_i = 3$ and the categories labelled 1, 2, and 3, we have

$$K_0(x_i - y_i) = \begin{cases} \lambda_1, & \text{if } x_i - y_i = 0 \\ \lambda_2, & \text{if } x_i - y_i = 1 \\ \lambda_3, & \text{if } x_i - y_i = 2 \end{cases}$$

where $\lambda_1 > \lambda_2 > \lambda_3$ and $\lambda_1 + \lambda_2 + \lambda_3 = 1$.

The similarity of this discrete variable kernel method to the continuous variable case makes it exceptionally easy to use the kernel method when there is a mixture of variable types, a situation which often arises and can cause difficulties (see Section 5.4.2). Titterington (1977) has extended the method to handle incomplete observation vectors and this is discussed in Chapter 8.

As an example of this method consider the General Health Questionnaire (GHQ). This is a psychiatric screening instrument used for identifying people likely to be suffering from non-psychotic psychiatric illnesses. One form of it consists of 30 questions, the answers to which are coded as 30 binary variables. For the purposes of illustration the available set of 97 known non-cases was divided into a design set of 49 and a test set of 48 and the available set of 82 known cases was divided into a design set of 39 and a test set of 43.

Table 5.1 Classifications resulting when the discrete kernel method was applied to scores on 30 binary variables from a psychiatric illness screening questionnaire: (a) The design set of 49 non-cases and 39 cases; (b) The test set of 48 non-cases and 43 cases

(a)

		True class	
		Non-case	Case
Predicted class	Non-case	43	13
	Case	6	26

(b)

		True class	
Predicted class	Non-case	44	19
	Case	4	24

Application of the jackknife method to estimate the λ's gave for the non-cases $\lambda = 0.96$ and for the cases $\lambda = 0.85$. In the general population the prevalence of such illness is thought to be of the order of 20 per cent, so the prior probability of the case class was set to 0.2 and the non-case class to 0.8. The kernel method then gave the classification results of Table 5.1—a misclassification rate of 22 per cent on the design set and 25 per cent on the test set. As usual, if one felt that a more equitable distribution of misclassifications between the two classes was desirable, then one could weight the pdf estimates by appropriate costs (see Chapter 1).

5.2.3 Nearest-neighbour methods

Just as the continuous variable kernel method is easily extended to categorical variables so also are the various nearest-neighbour methods. This should occasion no surprise in the light of the underlying similarities between the two methods already outlined in Chapter 2. The nearest-neighbour estimate was introduced in Section 2.4 in terms of the volume V of the hypersphere centred at $\mathbf{x}$, the point where the pdf estimate is desired, and with radius equal to the distance from $\mathbf{x}$ to its kth nearest neighbour amongst the design set points. For a class with n design set points the estimator is

$$\hat{p}(\mathbf{x}) = k/nV$$

This idea can be immediately extended to the categorical variable case. There is, however, a slight difficulty. To illustrate, suppose all variables are binary and let the distance between two points be the number of components in which they differ. Then there are $\binom{d}{i}$ cells at distance i from any given cell. If the kth nearest neighbour to $\mathbf{x}$ is at distance i from $\mathbf{x}$ then obviously all cells closer than i to $\mathbf{x}$ should be included in V. Similarly, all cells further than i from $\mathbf{x}$ should be excluded. But what about those cells at exactly distance i from $\mathbf{x}$? Clearly the answer is to include a proportion of them. Thus, let n_i points fall at distance i from $\mathbf{x}(i = 0, 1, \ldots, d)$ and suppose that the kth point is at distance j from $\mathbf{x}$. That is

$$n_{j-1} < k \leq n_j$$

Then

$$V = \binom{d}{0} + \binom{d}{1} + \cdots + \binom{d}{i-1} + \left(\frac{k - n_{i-1}}{n_i - n_{i-1}}\right)\binom{d}{i}$$

The extensions allowing explicit pdf estimation to be avoided (for example, by seeing which class has a majority amongst the k nearest neighbours) discussed in Section 2.4 can also be adapted for categorical variables.

It should be noted that Hills (1967) introduces a discrete variable probability function estimation method which he calls nearest neighbour. However, since this method fixes a volume and determines the proportion of points falling in that volume it is really a discrete variable kernel method with a kernel which is uniform within a certain range and zero outside it.

5.3 PARAMETERIZATIONS OF THE PROBABILITY FUNCTIONS

Again paralleling the continuous variables case many attempts have been made to approximate discrete probability functions using orthogonal series expansions. Most of this work has concentrated on the special case of binary variables.

We have already commented that in an ideal case one could approximate the overall multinomial distribution by estimating the probability in each cell, but this is seldom possible. Such an approach makes no assumptions about the forms of the class-conditional probability functions and requires the estimation of $2^d - 1$ independent parameters (for binary variables). If we use an orthogonal series method and similarly make no restrictive assumptions on the form of the probability functions then, although we have reparameterized these functions so that a different set of parameters needs to be estimated, there are still $2^d - 1$ of them. The methods of the preceding section solved the problem of reducing the number of degrees of freedom from $2^d - 1$ by imposing constraints on the cells—the fundamental smoothing of kernel methods assumes that the probability estimates in nearly cells are highly correlated, while k-NN methods assume that all cells within a certain distance from $\mathbf{x}$ have the same probability value. The methods of this section solve the problem by making simplifying assumptions about the form of the probability functions. In general these assumptions are that the higher order terms of the expansions are negligible, or at least may be ignored. Quite what interpretation to put on this depends on the precise form of the expansion, as will become apparent below. A particularly simple result can be obtained by assuming that the variables are independent and several expansions give this model if all but the first-order terms are deleted. We begin by demonstrating this assumption for logarithmic models.

5.3.1 Logarithmic models

A series of popular approaches to binary discrimination problems have been based on logarithmic transformations. A simple one is obtained if we assume that each class-conditional probability function can be written as

$$P(\mathbf{x}\,|\,\omega_i) = \exp(\boldsymbol{\alpha}_i'\mathbf{X})$$

i.e. the logarithms of the class-conditional functions are linear. $\mathbf{X}$ here can be $\mathbf{x}$ or, more usually, $\mathbf{X}' = (1, \mathbf{x}')$ to allow for a constant term. This case is discussed below. More generally, however, we can take higher order terms, leading ultimately to the saturated model, involving no restrictive assumptions at all, with

$$\mathbf{X}' = (1, x_1, \ldots, x_d, x_1x_2, \ldots, x_1x_d, x_2x_3, x_2x_4, \ldots, \ldots, x_1x_2x_3 \ldots x_d)$$

and where the vector of coefficients

$$\boldsymbol{\alpha}_i = (\alpha_{i0}, \alpha_{i1}, \alpha_{i2}, \ldots, \alpha_{i,\,2^d-1})$$

is to be determined from the design set. (Normally, however, if one wanted a saturated model one would use the multinomial approach discussed in Section 5.2.1.)

Bayes's theorem gives

$$\begin{aligned} P(\omega_i|\mathbf{x}) &= P(\mathbf{x}\,|\,\omega_i)P(\omega_i)/P(\mathbf{x}) \\ &= \mathrm{e}^{\boldsymbol{\alpha}_i'\mathbf{X}}P(\omega_i)/P(\mathbf{x}) \end{aligned}$$

so that a decision rule is

$$\ln\frac{P(\omega_i|\mathbf{x})}{P(\omega_j|\mathbf{x})} = (\boldsymbol{\alpha}_i' - \boldsymbol{\alpha}_j')\mathbf{X} + \ln\frac{P(\omega_i)}{P(\omega_j)} > 0, \quad \forall j \neq i \Rightarrow \mathbf{x} \in \Omega_i \qquad (1)$$

Taking $\mathbf{X}' = (1, \mathbf{x}')$ is equivalent to assuming that the variables are independent for each class and is an interesting special case. Thus, let $P_{ik} = P(x_k = 1\,|\,\omega_i)$. Then, under the independence assumption

$$P(\mathbf{x}\,|\,\omega_i) = \prod_{k=1}^{d} P_{ik}^{x_k}(1 - P_{ik})^{1-x_k}$$

giving

$$\ln P(\mathbf{x}|\omega_i) = \sum_{k=1}^{d} [x_k \ln P_{ik} + (1 - x_k)\ln(1 - P_{ik})]$$

$$= \sum_{k=1}^{d} x_k \ln \frac{P_{ik}}{1 - P_{ik}} + \sum_{k=1}^{d} \ln(1 - P_{ik})$$

Here, then

$$\alpha_{ik} = \ln \frac{P_{ik}}{1 - P_{ik}}$$

where the subscript i of α_{ik} refers to the class and k to the position in vector $\boldsymbol{\alpha}_i$, and

$$\alpha_{i0} = \sum_{k=1}^{d} \ln(1 - P_{ik})$$

The decision rule (1) thus becomes

$$\sum_{k=1}^{d} x_k \ln \frac{P_{ik}(1 - P_{jk})}{P_{jk}(1 - P_{ik})} + \sum_{k=1}^{d} \ln \frac{1 - P_{ik}}{1 - P_{jk}} + \ln \frac{P(\omega_i)}{P(\omega_j)} > 0, \quad \forall j \neq i \Rightarrow \mathbf{x} \in \Omega_i$$

Returning to the general case (where an independence assumption is not justified) we can find convenient estimators for the $\boldsymbol{\alpha}_i$ if we use

$$\begin{aligned} X_0 &= 1 \\ X_1 &= 2x_1 - 1 \\ X_2 &= 2x_2 - 1 \\ &\vdots \\ X_d &= 2x_d - 1 \\ X_{d+1} &= (2x_1 - 1)(2x_2 - 1) \\ &\vdots \\ X_{2^d-1} &= (2x_1 - 1) \ldots (2x_d - 1) \end{aligned}$$

These X_i are in fact orthogonal polynomials since

$$\sum_{\mathbf{x}} X_i X_j = \begin{cases} 2^d, & i = j \\ 0, & i \neq j \end{cases}$$

where the summation is over all 2^d cells. They are called *Rademacher–Walsh* polynomials. Expanding $\ln P(\mathbf{x}|\omega_i)$ in terms of these polynomials it is not difficult to show by usual series expansion methods (cf. the discussion of Fourier series estimation of continuous pdfs in Chapter 2) that an estimate $\hat{\alpha}_{ik}$

of α_{ik} is given by

$$\hat{\alpha}_{ik} = \frac{1}{n} \sum_{j=1}^{n} \frac{1}{2^d} X_k(\mathbf{x}_j)$$

An approximation of any desired degree of accuracy can be obtained by truncating the series appropriately and, using the properties of orthogonal series expansions, we can show that the $\hat{\alpha}_{ik}$ given above minimize the mean squared error between the experimental realization of $\ln P(\mathbf{x} | \omega_i)$ and its estimate in terms of the basis $\{X_k\}$. That said, however, it should be noted that for all except very small d too many parameters will be needed unless $\mathbf{X}' = (1, \mathbf{x}')$. Not only does a large number of parameters imply practical estimation difficulties (cf. the multinomial method) but it also implies a great risk of overfitting—a very good fit may be obtained on the original classified sample, but it may not generalize well to new observations. In an effort to alleviate this difficulty of large numbers of parameters Anderson (1975) has suggested a way of reducing the required number in a quadratic logistic model by making suitable approximations.

Note that all the estimates $\hat{P}(\mathbf{x} | \omega_i)$ are non-negative (in fact, positive). This is not true if we expand $P(\mathbf{x} | \omega_i)$ directly instead of $\ln P(\mathbf{x} | \omega_i)$. This is discussed in Section 5.3.2.

So far we have assumed that the logarithm of each of the class-conditional probability functions could be adequately represented by a linear function. Slightly more generally we can assume only that the *difference* between the logarithms of the class-conditional probability functions is linear, making no linearity assumptions about the class-conditional probability functions themselves. (Assuming the logarithm of each class-conditional function is linear implies linearity of the difference, but linearity of the difference does not imply linearity of each function separately.) This more general assumption is equivalent to assuming that the coefficients of all terms not explicitly included in $\mathbf{X}$ are identical in the two populations (earlier we assumed such coefficients to be zero). Anderson (1972) adopts this approach, except that he assumes that

$$\ln P(\omega_i | \mathbf{x}) - \ln P(\omega_j | \mathbf{x})$$

rather than

$$\ln P(\mathbf{x} | \omega_i) - \ln P(\mathbf{x} | \omega_j)$$

is linear. (Note that these two assumptions of linearity are equivalent, the only difference between the expansions being a constant term involving $P(\omega_i)$ and $P(\omega_j.)$ Thus

$$\ln P(\omega_i | \mathbf{x}) / P(\omega_N | \mathbf{x}) = \boldsymbol{\alpha}_i' \mathbf{X}, \qquad i = 1, \ldots, N - 1 \tag{2}$$

From this discussion above it is clear that with $\mathbf{X}' = (1, \mathbf{x}')$ this model is exact when the x_i are independent binary variables. It is also exact if the class-conditional pdfs are multivariate normal with identical variance–covariance matrices.

Anderson uses the maximum likelihood method to obtain a set of equations which can be solved for the $\boldsymbol{\alpha}_i$. An iterative Newton–Raphson method to solve these equations is as follows: Let

$$f_{ij} = \sum_{\mathbf{x}} [n_{i\mathbf{x}} - n_{\mathbf{x}} P(\omega_i | \mathbf{x})] X_j$$

where $n_{i\mathbf{x}}$ is the number of points from class i in cell $\mathbf{x}$, $n_{\mathbf{x}} = \sum_i n_{i\mathbf{x}}$, where i ranges from 1 to $N - 1$, and j ranges from 0 to m (over the components of $\mathbf{X}$). Then, using equation (2) and

$$P(\omega_N | \mathbf{x}) = \left\{ 1 + \sum_{i=1}^{N-1} e^{\boldsymbol{\alpha}_i' \mathbf{x}} \right\}^{-1}$$

gives

$$\frac{\partial f_{ij}}{\partial \alpha_{tl}} = \sum_{\mathbf{x}} n_{\mathbf{x}} P(\omega_i | \mathbf{x}) P(\omega_t | \mathbf{x}) X_j X_l \qquad (i \neq t)$$

and

$$\frac{\partial f_{ij}}{\partial \alpha_{il}} = -\sum_{\mathbf{x}} n_{\mathbf{x}} P(\omega_i | \mathbf{x})(1 - P(\omega_i | \mathbf{x})) X_j X_l$$

Now let $\mathbf{F}$ be the $(N - 1)(m + 1) \times (N - 1)(m + 1)$ matrix with elements $F_{ij,tl} = \partial f_{ij} / \partial \alpha_{tl}$ where the order of the pairs (i, j) and (t, l) is $(1, 0)$, $(1, 1)$, ... $(1, m)$, $(2, 0)$ . . $(N - 1, m)$ and let $\mathbf{f}$ be the column vector with elements f_{ij} in the same order. Then if $\mathbf{f}_a$ and $\mathbf{F}_a$ are $\mathbf{f}$ and $\mathbf{F}$ when $\mathbf{a}' = (\boldsymbol{\alpha}_1, \ldots . \boldsymbol{\alpha}_{N-1})$ the iterative scheme is

$$\mathbf{a}_1 = \mathbf{a}_0 - \mathbf{F}_{\mathbf{a}_0}^{-1} . \mathbf{f}_{\mathbf{a}_0}$$

Anderson suggests that it is not necessary to update $\mathbf{F}^{-1}$ at every step of the iteration.

Anderson and Richardson (1979) extend this work on logistic discrimination by reducing the bias in the maximum likelihood estimates of the parameters, and consequently reducing the bias in the linear discriminant function.

Berkson (1955) has also investigated another way of estimating the $\boldsymbol{\alpha}_i$. He finds $\hat{\boldsymbol{\alpha}}$ to minimize the logit χ^2 defined by

$$\sum_{\mathbf{x}} \frac{n_{1\mathbf{x}} n_{2\mathbf{x}}}{n_{1\mathbf{x}} + n_{2\mathbf{x}}} \left[\ln \frac{n_{1\mathbf{x}}}{n_{2\mathbf{x}}} - \boldsymbol{\alpha}' \mathbf{X} \right]^2$$

Since problems will be encountered if some $n_{i\mathbf{x}} = 0$ Berkson recommends substituting $n_{i\mathbf{x}} = \frac{1}{2}$ for empty cells.

During the last few years considerable developments have been made in the analysis of contingency tables using log-linear models (see, for example, Bishop, Fienberg, and Holland, 1975, or, for a simpler introduction, Fienberg, 1977). Such models represent the probability distribution by an expansion

$$P(\mathbf{x}) = \exp(\alpha_0 + \alpha_1(x_1) + \alpha_2(x_2) + \ldots + \alpha_d(x_d) + \alpha_{12}(x_1, x_2) + \ldots + \alpha_{1,2,\ldots,d}(x_1, \ldots, x_d))$$

The $\alpha_{i,\ldots,j}$ parameters are defined as follows

$$\alpha_0 = \sum_{\mathbf{x}} \ln \hat{P}(\mathbf{x}) \Big/ \prod_{i=1}^{d} g_i$$

where g_i is the number of categories for the ith variable

$$\alpha_1(a) = \sum_{\substack{\mathbf{x} \\ x_1=a}} \ln \hat{P}(\mathbf{x}) \Big/ \prod_{i=2}^{d} g_i - \alpha_0$$

$$\vdots$$

$$\alpha_d(a) = \sum_{\substack{\mathbf{x} \\ x_d=a}} \ln \hat{P}(\mathbf{x}) \Big/ \prod_{i=1}^{d-1} g_i - \alpha_0$$

$$\alpha_{12}(a, b) = \sum_{\substack{\mathbf{x} \\ x_1=a, x_2=b}} \ln \hat{P}(\mathbf{x}) \Big/ \prod_{i=3}^{d} g_i - \alpha_2 - \alpha_1 - \alpha_0$$

$$\vdots$$

$$\alpha_{1\ldots d}(a, \ldots, c) = \sum_{\substack{\mathbf{x} \\ \mathbf{x}=(a,\ldots,c)}} \ln \hat{P}(\mathbf{x}) - \ldots - \alpha_0$$

Where $\hat{P}(\mathbf{x})$ is the expected value of $P(\mathbf{x})$ under the model assumptions. It is instructive to view these as main effects and interaction terms analogous to those of analysis of variance. For example, if $d = 2$ and both variables are binary, α_0 becomes

$$\alpha_0 = \frac{\ln \hat{P}((0, 0)) + \ln \hat{P}((0, 1)) + \ln \hat{P}((1, 0)) + \ln \hat{P}((1, 1))}{4}$$

and α_1 becomes

$$\alpha_1 = \frac{\ln \hat{P}((0, 0)) + \ln \hat{P}((0, 1))}{2} - \alpha_0$$

$$= \frac{\ln \hat{P}((0, 0)) + \ln \hat{P}((0, 1)) - \ln \hat{P}((1, 0)) - \ln \hat{P}((1, 1))}{4}$$

which are directly analogous to the mean and first main effect constrasts of analysis of variance. Rather than a Procrustean truncation of all high order interactions, however, expected frequencies arising from the models are estimated and a model is selected by goodness of fit tests. Common criteria for such selections are the Pearson χ^2 statistic

$$\chi_p^2 = \sum_{\mathbf{x}} [O(\mathbf{x}) - E(\mathbf{x})]^2 / E(\mathbf{x})$$

(where $O(\mathbf{x})$ is the observed number in cell $\mathbf{x}$ and $E(\mathbf{x})$ is the expected number in cell $\mathbf{x}$), and the likelihood ratio χ^2 statistic

$$\chi^2_{\mathrm{L}} = 2 \sum_{\mathbf{x}} O(\mathbf{x}) \ln[O(\mathbf{x})/E(\mathbf{x})]$$

Under the null hypothesis that the model is correct (and assuming large samples) χ^2_{P} and χ^2_{L} are approximately distributed as χ^2 with degrees of freedom given by

$$\text{(total number of cells)} - \text{(number of parameters fitted)}$$

We can thus a test a particular model to see if it provides a satisfactory fit. The usual approach, however, is to choose a model by a sequential procedure, adding in or subtracting out parameters until an adequate fit is achieved with a sufficiently simple model. Since the associated χ^2_{L} or χ^2_{P} statistics will not be independent we cannot simply take them as providing valid tests of the models in such a sequential procedure. Several methods have been suggested to alleviate this problem, and the reader is referred to Bishop *et al*. (1975) and Fienberg (1977) for outlines and guides to the literature. One general technique which is valid when comparing two models A and B, where B is a special case of A (in the sense that A includes all of the B parameters, and more besides) is to test the significance of the additional parameters by the statistic

$$\chi^2_{\mathrm{L}}(B) - \chi^2_{\mathrm{L}}(A)$$

taking it as approximately χ^2_{L} distributed with

$$\mathrm{df}\,(B) - \mathrm{df}\,(A)$$

degrees of freedom.

To illustrate the method consider a study of satisfaction with the local environment. One of the questions asked was: 'How satisfied are you with this area as a place to live in?'. The responses were recoded to binary form giving two classes: satisfied and dissatisfied. Six questions were used as predictor variables, whether or not the respondent agreed that:

(1) There were enough shops within easy reach.
(2) The local bus service was good.
(3) The road outside was difficult to cross.
(4) There were plenty of places of entertainment within easy reach.
(5) People in the neighbourhood were unfriendly.
(6) There were enough parks within easy reach.

With each of these predictor variables coded as binary there are 64 parameters in each of the two classes, 128 in all.

When log-linear models were fitted, sequentially eliminating terms by the procedure described above, adequate fits were still obtained with only the main effects and interactions shown in Table 5.2. (For the 'satisfied' class the likelihood ratio χ^2_{L} was 54.69 with 52 degrees of freedom, giving $p < 0.373$, not sufficiently small for us to reject the hypothesis that the model fits the

Table 5.2 Main effects and interactions necessary for adequate log-linear approximations to the distributions of people satisfied and dissatisfied with their local environment

Satisfied	Dissatisfied
1	1
2	2
3	3
4	4
5	5
6	6
1 × 2	1 × 2
1 × 4	1 × 3
1 × 6	1 × 4
2 × 4	1 × 6
4 × 6	2 × 4
	2 × 6
	3 × 6
	4 × 6
	1 × 2 × 4
	1 × 3 × 6
	2 × 4 × 6

Note: Key to numbers is in text.

data. For the 'dissatisfied' class $\chi_2^2 = 55.02$ with 46 degrees of freedom giving $p < 0.17$.) The distributions are thus now described by 30 instead of 128 parameters. (There are only 28 in Table 5.2, the remaining two being due to the absolute sizes of the two classes, i.e. the number of observations in each class.) The probability density functions estimated from these simplified descriptions of the data led to a misclassification rate of 34 per cent compared to a misclassification rate of 30 per cent when the full multinomial model was used.

Note that an alternative way to proceed would be to treat the entire 7-way table as a single probability function.

5.3.2 Other series methods

In the preceding section we discussed how $\ln \mathrm{P}(\mathbf{x}|\omega_i)$ and $\ln P(\omega_i|\mathbf{x})/P(\omega_j|\mathbf{x})$ could be estimated by a series of functions. There is, however, no reason why we should not try expanding other functions of $P(\mathbf{x}|\omega_i)$ instead of the logarithm. In particular, we could expand $P(\mathbf{x}|\omega_i)$ itself. This is effectively what Ott and Kronmal (1976) have done, writing

$$\hat{P}(\mathbf{x}) = \frac{1}{2^{d+1}} \sum_{\mathbf{r}} \hat{\alpha}_{\mathbf{r}} X_{\mathbf{r}}(\mathbf{x}) \tag{3}$$

where $\mathbf{r}$ is a $(d+1)$ dimensional binary vector (the $(d+1)$st dimension distinguishing between the two classes), the summation ranges over all possible values, $\mathbf{r}$, $\hat{\alpha}_{\mathbf{r}} = \Sigma_{\mathbf{x}} X_{\mathbf{r}}(\mathbf{x}) n_{i\mathbf{x}}/n$, and $\{X_{\mathbf{r}}(\mathbf{x}) = (-1)^{\mathbf{x}'\mathbf{r}}\}$ is an orthogonal set of functions. Rather than simply truncating the series Ott and Kronmal present four methods for selecting terms.

(i) Basic Fourier Method: Using as a criterion the mean summed squared error (MSSE)

$$E \sum_{\mathbf{x}} \sum_{i=1}^{2} [\hat{P}(\mathbf{x}|\omega_i) - P(\mathbf{x}|\omega_i)]^2$$

it can be shown that an increase in accuracy will result by including a term $\hat{\alpha}_{\mathbf{r}} X_{\mathbf{r}}$ in the expansion (3) if

$$\hat{\alpha}_{\mathbf{r}}^2 > 2/(n+1)$$

(ii) Fourier procedure with unbiased difference estimate: Instead of expanding $P(\mathbf{x}|\omega_i)$ separately, as (3) effectively does, expand $P(\mathbf{x}|\omega_1) - P(\mathbf{x}|\omega_2)$ as

$$\frac{1}{2^d} \sum_{\mathbf{r}} \beta_{\mathbf{r}} X_{\mathbf{r}}(\mathbf{x})$$

and use $\hat{\beta}_{\mathbf{r}} = \Sigma_{\mathbf{x}} X_{\mathbf{r}}(\mathbf{x})(n_{\mathbf{x}1} - n_{\mathbf{x}2})/n$. $\mathbf{r}$ here is a d-dimensional binary vector. As in (i) terms with coefficient $\hat{\beta}_{\mathbf{r}}^2 < 2/(n+1)$ are not included in the expansion in order to minimize the MSSE.

(iii) Fourier procedure with maximum likelihood estimate of the difference: Here the biased maximum likelihood estimate $\beta_{\mathbf{r}}^{*2}$ of $\beta_{\mathbf{r}}^2$ replaces $\hat{\beta}_{\mathbf{r}}^2$ in (ii). The variance of $\beta_{\mathbf{r}}^{*2}$ is smaller by a factor of $n^2/(n-1)^2$ than that of $\beta_{\mathbf{r}}^2$. The selection rule for this method is: set $\beta_{\mathbf{r}}^*$ to zero unless

$$\beta_{\mathbf{r}}^* > 1/(n+1)$$

(iv) Fourier procedure with weighted coefficients: Instead of a sharp 'inclusion/ exclusion' decision for each term, all terms are included but with a weight $\mathrm{w}_{\mathbf{r}}$ for term $\beta_{\mathbf{r}}^* X_{\mathbf{r}}(\mathbf{x})$. The MSSE is minimized if

$$\hat{\mathrm{w}}_{\mathbf{r}} = n\beta_{\mathbf{r}}^{*2}/(1 + (n-1)\beta_{\mathbf{r}}^{*2})$$

Ott and Kronmal carried out an extensive simulation study on the performance of these methods, as well as comparing them with the complete multinomial method, the logistic method based on $\ln P(\mathbf{x}|\omega_1)/P(\mathbf{x}|\omega_2)$, and a method assuming the variables were independent. When the methods were ranked according to their mean error rates the independence method came top for samples of size 100, but for larger samples the Fourier method with weighted coefficients did best. Of course, simulation results depend on the populations that the methods are tested on.

The Rademacher–Walsh polynomials used in the preceding section to expand $\ln P(\mathbf{x}|\omega_i)$ can be used to expand $P(\mathbf{x}|\omega_i)$ directly as Ott and

Kronmal have done. In general, however, there are two possible disadvantages to the direct expansion of $P(\mathbf{x}|\omega_i)$. The first is that estimates of $P(\mathbf{x}|\omega_i)$ can lie outside the range (0, 1). Of course, since classification, based on comparisons of $\hat{P}(\mathbf{x}|\omega_i)$ with $\hat{P}(\mathbf{x}|\omega_j)$, is the ultimate aim, this may not be a disadvantage—we can compare the estimates even if they are negative. The second problem is that the independence model does not result if only the first-order terms are taken. Independence of two marginals implies, of course, that the joint distribution may be written as a product of the two variables in question. That is to say, for independence to be the simplest special case we would need to fit a multiplicative rather than an additive model (and this is precisely what happens if we first take a log transform). Bahadur (1961) has attempted to resolve this difficulty for the additive model by expanding $P(\mathbf{x}|\omega_i)$ in the form

$$P(\mathbf{x}|\omega_i) = P_1(\mathbf{x}|\omega_i)\Bigg[1 + \sum_{j<k} \rho_{jk}u_j u_k + \sum_{j<k<l} \rho_{jkl}u_j u_k u_l + \ldots + \rho_{12\ldots d}u_1 u_2 \ldots u_d\Bigg] \tag{4}$$

where

$$u_j = (x_j - P_j)/\sqrt{P_j(1-P_j)}, \qquad P_j = P(x_j = 1|\omega_i)$$

and

$$P_1(\mathbf{x}|\omega_i) = \prod_{j=1}^{d} P_j^{x_j}(1-P_j)^{1-x_j}$$

That is, $P(\mathbf{x}|\omega_i)$ is split into two factors, the first accounting for as much of $P(\mathbf{x}|\omega_i)$ as can be explained by assuming independence and the second being a polynomial expansion of the correction term between $P(\mathbf{x}|\omega_i)$ and $P_1(\mathbf{x}|\omega_i)$.

In this expansion the (correlation) coefficients $\rho_{jk\ldots m}$ can be estimated from

$$\hat{\rho}_{jk\ldots m} = \frac{1}{n}\sum_{1}^{n} u_j u_k \ldots u_m$$

the summation being over the n sample points.

The easiest way to derive this series is to define polynomials $X_0 = 1$, $X_1 = u_1, \ldots, X_d = u_d, X_{d+1} = u_1 u_2, \ldots, X_{2^d-1} = u_1 u_2 u_3 \ldots u_d$. These satisfy the orthogonality condition

$$\sum_{\mathbf{x}} X_j X_k P_1(\mathbf{x}|\omega_i) = \begin{cases} 1, & \text{if } j = k \\ 0, & \text{if } j \neq k \end{cases}$$

and one can apply usual methods of orthogonal series expansion to $P(\mathbf{x}|\omega_i)/P_1(\mathbf{x}|\omega_i)$.

Brunk and Pierce (1974) have investigated a similar approach using, as the second factor in (4), a term of the form $\exp(\boldsymbol{\alpha}_i'\mathbf{X})$.

Martin and Bradley (1972) expand yet another function of $P(\mathbf{x}|\omega_i)$, namely the normalized difference between $P(\mathbf{x}|\omega_i)$ and $P(\mathbf{x})$ (where $P(\mathbf{x}) = \Sigma_i P(\omega_i)P(\mathbf{x}|\omega_i)$). Thus

$$\frac{P(\mathbf{x}|\omega_i) - P(\mathbf{x})}{P(\mathbf{x})} = \boldsymbol{\alpha}_i'\mathbf{X}(\mathbf{x})$$

They use the Rademacher–Walsh polynomials discussed above as the components of $\mathbf{X}$ and obtain

$$\alpha_{ij} = \frac{1}{2^d} \sum_{\mathbf{x}} X_j(\mathbf{x}) \frac{P(\mathbf{x}|\omega_i) - P(\mathbf{x})}{P(\mathbf{x})}$$

(conditional on $P(\mathbf{x}) \neq 0$), substituting the maximum likelihood estimates $\hat{P}(\mathbf{x}|\omega_i) = n_{i\mathbf{x}}/n_i$ and $\hat{P}(\mathbf{x}) = \Sigma P(\omega_i)\hat{P}(\mathbf{x}|\omega_i)$ (with $P(\omega_i)$ assumed known) to give estimates $\hat{\alpha}_{ij}$.

5.4 OTHER ASPECTS OF DISCRETE VARIABLES

5.4.1 Fisher's linear discriminant function

In Section 4.5 we showed how Fisher's linear discriminant function (LDF) arises from an attempt to maximize the ratio of between-class scatter to the within-class scatter. This maximization and the resulting LDF do not require any distributional assumptions. However, when looked at from a likelihood ratio point of view Fisher's LDF is seen to be optimal when the two classes have normal distributions with equal covariance matrices. If the classes are not so distributed the method is no longer optimal. Despite this, the widespread availability of Fisher's LDF in statistical program packages has led to its being applied to data which are not known to follow the required distributional forms. In view of this several authors have investigated the performance of Fisher's LDF with non-normal classes (see, for example, Gilbert, 1968, Moore, 1973, and Krzanowski, 1977). For binary variables the general conclusions seem to be that if the true decision surface is roughly linear, Fisher's LDF will perform satisfactorily, but if the true surface is markedly non-linear (e.g. with $d = 2$ if $P(00|\omega_1) > P(00|\omega_2)$ and $P(11|\omega_1) > P(11|\omega_2)$ but $P(01|\omega_1) < P(01|\omega_2)$ and $P(10|\omega_1) < P(10|\omega_2)$) then it will perform badly. However, it should be noticed that if a non-linear decision surface is discovered it might be possible to transform the data. As an example consider Table 5.3(a), where upper case letters signify larger numbers than the matching lower case letters. Table 5.3(a) would require a non-linear decision surface. In Table 5.3(b) we have replaced x_2 by a new variable x_2' defined as

$$x_2' = 0, \quad \text{if } x_1 \text{ and } x_2 \text{ are the same}$$
$$x_2' = 1, \quad \text{if } x_1 \text{ and } x_2 \text{ are different}$$

Now a linear decision surface would suffice. Bloomfield (1974) studies such ways of recoding multivariate binary data so that the new variables permit the

Table 5.3 Transforming multivariate binary data so that linear decision surfaces are effective. Capital letters signify larger numbers than the matching lower case letters

(a)

Class 1

x_1 \ x_2	0	1
0	A	b
1	c	D

Class 2

x_1 \ x_2	0	1
0	a	B
1	C	d

(b)

Class 1

x_1 \ x_2'	0	1
0	A	b
1	D	c

Class 2

x_1 \ x_2'	0	1
0	a	B
1	d	C

data structure to be described by a simple model (such as by assuming independence between the variables, or a low order log-linear model). Krzanowski (1979) extends this to mixtures of binary and continuous variables and suggests transformations which may be applied with the aim of improving Fisher's linear discriminant function.

In spite of all this one should be aware that while Fisher's linear discriminant function may yield satisfactory results on categorical data, it certainly need not always do so. To illustrate, during the course of the enuresis study outlined in Exercise 5.6, a comparison was made of those who failed to respond to treatment and those who relapsed after an apparent cure. Subsequently, observations on a further 33 patients became available. The linear discriminant function misclassified 9 (27 per cent) of these while the kernel method misclassified only 1 (3 per cent). (A different variable set was used in this part of the study from those described in Exercise 5.6.)

5.4.2 Mixtures of variable types

In certain fields (such as medicine) it is common to find that some of the variables are continuous while others are discrete. In the past the most popular approaches to such problems seem to have been to treat the discrete variables as continuous and to apply a straightforward linear decision surface—usually Fisher's method—or to group the continuous variables into categories (see, for example, Cochran and Hopkins, 1961). Such approaches have obvious associated disadvantages. More recent approaches include that of Krzanowski (1975) who estimates separate continuous variable decision surfaces in each cell generated by the product of the discrete variables. As mentioned above, Krzanowski (1979) suggests transformations which can improve the performance of Fisher's linear discriminant function on a mixture of variable types.

We have already discussed separately the case of class-conditional probability functions which have independent binary variables, and the case of class-conditional pdfs following multivariate normal distributions with equal variance–covariance matrices. In both of these cases an exact model is given by Anderson's (1972) logistic method

$$\ln P(\omega_i|\mathbf{x}) - \ln P(\omega_j|\mathbf{x}) = \boldsymbol{\alpha}'\mathbf{x}$$

Such a model is also exact if some of the variables follow a multivariate normal distribution (with equal variance–covariance matrices across classes) and the remainder are independent binary variables.

Aitchison and Aitken (1976) have suggested a flexible non-parametric method based on the general kernel estimator formed by combining the ideas of Section 1.3 with those of Section 5.2.2. We thus have

$$\hat{p}(\mathbf{x}, \mathbf{y}|\omega_i) = \frac{1}{n}\sum_{j=1}^{n} K(\mathbf{x}|\mathbf{u}_j, h)B(\mathbf{y}|\mathbf{v}_j\lambda)$$

where $\mathbf{x}$ and $\mathbf{u}_j$ are vectors of continuous variables, $\mathbf{y}$ and $\mathbf{v}_j$ are vectors of discrete variables, K is the kernel for the continuous variables, B is the kernel for the binary variables, and $(\mathbf{u}_j', \mathbf{v}_j')$ is the jth point of the design set. Similar extensions of the k-nearest-neighbour method are also possible.

5.5 CHOICE OF METHOD

Given the number of different ways of approaching discriminant analysis on categorical variables the question naturally arises of which method to choose. As usual there is no global best and one should be aware of the advantages and disadvantages of the various methods so that an appropriate choice can be made for any particular problem. Table 5.4 lists some of these relevant properties of the methods. In summary, the multinomial method makes no assumptions at all about the class-conditional distribution functions, but it is only feasible if d is small or n is very large. Non-parametric kernel and nearest-neighbour methods restrict the number of distributions that can be modelled by making implicit assumptions about the correlations between neighbouring cells. They in effect assume that nearby cells have similar probability values, whereas the multinomial method, which can be viewed as a limiting case as this restriction is relaxed, makes no such assumption. The parametric methods, whether obtained by expanding the logarithm of class-conditional distribution functions or by some other expansion, assume that these distributions can be adequately modelled by the chosen series. More flexibility, and hence a greater chance of having a good enough model, can be obtained by introducing further parameters—up to the limiting case of $2^d - 1$

Table 5.4 Properties of discrete variable discriminant analysis methods

Method	Advantages	Disadvantages
Multinomial	No assumptions	Small d or very large n
Kernel	Only assumption is high correlations between nearby cells	Limited if classification speed is important—though could be modified
k-Nearest neighbour	As kernel	As kernel
Simple logarithmic $\hat{p}(\mathbf{x}) = \exp(\alpha_0 + \boldsymbol{\alpha}'\mathbf{x})$	Only d parameters/class (in binary case)	Assumes independent variables
Higher order logarithmic	No independence assumption	More parameters to be estimated. Risk of overfitting
Logistic $\dfrac{\hat{p}_i(\mathbf{x})}{\hat{p}_j(\mathbf{x})} = \exp(\alpha_0 + \boldsymbol{\alpha}'\mathbf{x})$	More general than simple logarithmic. Independent variables is a special case	Usual assumption of parametric methods: that the probability function can be modelled adequately by the chosen form
Log-linear	Parameters can be interpreted in terms of main effects and interactions	Small d, large n
Direct expansions $\hat{p}(\mathbf{x}) = \alpha_0 + \boldsymbol{\alpha}'\mathbf{x}$	Easy to interpret	Independence not the simplest special case. Estimates can be outside (0, 1)
Hybrids	Independence is simplest special case	As logistic. Awkward interpretation?

independent parameters (in the binary case) which describes the observed distribution perfectly and is equivalent to the multinomial method. There are, however, estimation difficulties and dangers of overfitting the design set associated with having too many parameters. Frequently an acceptable compromise is reached by including only first-order terms in the series. If terms are to be selected by a model-fitting approach (as in the log-linear method), rather than by arbitrary truncation of higher order terms, then sufficient design set points must be available for reliable estimation of the parameters.

5.6 FURTHER READING

As far as the author is aware there is, at the time of writing, only one book solely devoted to discriminant analysis on categorical variables. This is that of Goldstein and Dillon (1978). The book is basically a survey of work on discrete discriminant analysis and acts as a useful source to the original references. In most cases, however, it does not develop the techniques it discusses and serious students will find it necessary to refer to the original publications. Apart from summarizing the methods which have been suggested, it also contains chapters on estimating error rates and selecting variables as well as some computer program listings (in FORTRAN).

Cox (1972) presents a concise summary of the analysis of multivariate data and is a useful, if brief overview. He does not, however, discuss the use of log-linear models; for this the book by Bishop, Fienberg, and Holland is recommended.

EXERCISES

5.1 Some approaches apply a multiplicative correction to an initial pdf estimate based on an assumption of independence of binary variables (see Section 5.3.2). One could also apply an additive correction. In fact, more symmetrically, one could use a mixture distribution for each class-conditional pdf. Devise an algorithm for estimating the parameters in the following model

$$p(\mathbf{x}|\omega_i) = \sum_{j=1}^{c} w_j \prod_{k=1}^{d} x_k^{p_{jk}}(1 - x_k)^{1-p_{jk}}$$

where

$$\sum_{j=1}^{c} w_j = 1; \qquad w_j > 0; \qquad p_{jk} \geqslant 0, \quad \text{for all } j, k$$

5.2 One way to alleviate the difficulties arising due to many empty cells (itself a result of the curse of dimensionality) would be to reduce the number of variables. The aim would be to do this without degrading the performance of the final discriminant functions, i.e. while retaining as much 'separability' between the class-conditional pdfs as possible. Discuss possible measures of separability, bearing in mind that an ideal one (such as misclassification rate) might not be feasible since a very large number of subsets of variables might have to be considered (see also Chapter 6).

5.3 Using standard computer packages for classical linear discriminant analysis (e.g. SPSS) and for fitting log-linear models (e.g. ECTA or GLIM, details are given in Everitt, 1977) compare the misclassification rates for the two approaches on several different sets of categorical data.

5.4 In some areas of behavioural research the constituent items of questionnaires are summed to provide an overall score. These scores are then compared with a threshold to classify respondents into one of two classes. By generating synthetic data sets of multivariate binary responses, investigate the advantages to be gained by weighting the individual items before summing them.

5.5 In some circumstances it is important that each classification be made as quickly as possible. This is the case, for example, in classifying the pixels of satellite photographs, where very large numbers must be classified each second so that estimates of (again, for example) overall crop

Table 5.5 Enuresis data (see Exercise 5.6)

V_1	V_2	V_3	V_4	V_5	V_6	V_7	Class	V_1	V_2	V_3	V_4	V_5	V_6	V_7	Class
0	0	0	0	0	1	0	3	1	0	0	0	0	1	0	3
1	0	0	0	0	1	0	3	0	0	0	0	0	1	0	3
0	0	0	0	0	1	0	3	1	0	0	0	0	1	1	3
0	0	0	0	0	0	0	3	0	0	0	0	0	1	0	2
0	0	0	0	0	1	0	3	0	0	0	0	0	1	0	3
0	0	0	0	0	1	0	2	1	0	0	0	1	1	1	1
1	0	0	0	0	1	0	2	0	0	0	0	1	1	0	3
0	0	0	0	0	1	1	2	0	0	0	0	0	1	0	2
1	0	0	0	0	1	1	1	0	0	0	0	1	1	0	1
1	0	0	0	0	1	0	2	1	0	0	0	1	1	1	1
1	0	0	0	0	1	0	2	0	0	1	0	1	1	0	3
0	0	0	0	0	1	0	3	0	0	0	0	1	1	1	3
1	0	0	0	0	0	0	1	0	0	0	0	1	1	0	3
1	0	0	0	1	1	0	2	0	0	0	0	1	1	1	3
1	0	0	0	0	1	0	3	1	0	0	0	0	1	0	1
1	0	0	0	0	1	0	3	1	0	0	0	1	1	1	3
0	0	0	0	1	1	0	2	0	0	0	0	1	1	0	3
0	0	0	0	1	1	0	3	0	0	0	0	1	1	0	3
0	0	0	0	1	1	0	3	0	0	0	0	1	1	0	2
1	0	0	0	0	0	0	1	0	0	0	0	1	1	1	3
0	0	0	0	0	1	0	2	0	0	0	0	1	0	0	2
0	0	0	0	1	1	0	3	1	0	0	0	1	1	0	3
1	0	0	0	0	1	0	1	1	0	0	0	1	1	1	3
1	0	0	0	0	1	0	3	1	0	0	0	1	1	0	2
1	1	0	0	0	1	1	2	0	0	0	0	1	1	0	3
0	0	0	0	0	0	0	3	1	0	1	0	1	1	1	2
1	0	0	0	1	0	0	3	0	0	1	0	1	0	0	3
1	0	0	0	0	1	0	2	0	0	0	0	1	1	0	2
1	1	1	1	0	1	1	2	1	1	0	0	1	1	0	2
0	0	0	0	1	1	0	2	0	0	0	0	1	1	0	3
0	0	0	0	1	1	0	3	1	0	0	0	1	1	0	2
1	0	0	0	1	1	0	2	1	0	0	0	1	1	0	2
0	0	0	0	1	1	0	3	0	0	0	0	1	1	0	3
1	0	0	0	1	1	0	2	1	0	1	0	1	1	1	3
0	0	0	0	0	1	1	3	0	0	0	0	1	0	0	3
0	0	0	0	1	1	0	2	0	0	0	0	1	1	0	1
1	0	0	0	0	1	0	2	1	0	0	0	1	1	1	1
1	0	0	0	1	1	0	3	0	0	0	0	1	1	0	2
1	0	0	0	1	1	0	3	0	0	0	0	1	1	0	3
1	0	0	0	1	1	0	3	0	0	0	0	1	1	1	2
0	0	1	0	1	1	0	3	0	0	0	0	1	1	1	3
1	0	0	0	1	1	0	2	1	0	0	0	1	1	0	3
0	0	0	0	0	1	0	1	0	0	0	0	1	1	0	3
0	0	0	0	0	1	0	1	0	0	0	0	1	1	0	3
1	0	0	0	0	1	1	3	1	0	0	0	1	1	1	2
0	0	0	0	0	1	0	2	0	0	0	0	1	1	0	3
1	0	0	0	1	1	1	1	1	0	0	0	1	1	0	3
1	0	0	0	1	0	1	1	1	0	0	0	1	1	1	1
0	0	0	0	0	1	0	3	1	0	0	0	1	1	0	2
0	0	0	0	0	1	0	3	0	0	0	0	1	0	0	3

Table 5.5 *continued*

V_1	V_2	V_3	V_4	V_5	V_6	V_7	Class	V_1	V_2	V_3	V_4	V_5	V_6	V_7	Class
1	0	0	0	1	1	0	2	1	0	0	1	1	1	0	1
1	0	0	0	1	1	0	2	1	0	1	0	1	0	1	3
0	0	0	0	1	1	0	3	1	0	0	0	1	1	0	2
0	0	0	0	0	1	0	3	0	0	0	0	1	1	0	2
0	0	0	0	0	1	0	3	1	0	0	0	1	0	0	1
1	0	0	0	0	1	0	1	1	0	0	0	1	1	0	2
0	0	0	0	1	1	0	3								

acreage can be made. Compare the speed of the kernel method with the logistic methods and the classical linear discriminant function method.

5.6 One method of treatment of enuretic children involves an alarm buzzer which wakes the child whenever the bed becomes wet. The data in Table 5.5 were collected in a study to investigate whether the outcome of treatment could be predicted from certain measurements. The criterion groups are 1 = fail, 2 = relapse after apparent cure, 3 = long-term cure. The variables are

v_1 Whether or not there were family background difficulties (1 = yes, 0 = no).
v_2 Urinary tract infection (1 = yes, 0 = no).
v_3 Whether wetting occurred during the day (1 = yes, 0 = no).
v_4 Whether soiling also occurred (1 = yes, 0 = no).
v_5 The child's age (0 if greater than 8, 1 if otherwise).
v_6 Whether the family had access to an inside w.c. (1 = yes, 0 = no).
v_7 Whether the child shared a room with more than one sibling (1 = yes, 0 = no).

Using this data set investigate how critical is the choice of λ when applying the kernel method. (I am indebted to Dr Sylvia Dische for permission to reproduce these data.)

CHAPTER 6
Variable Selection

6.1 INTRODUCTION

For any given classification problem there is an unlimited number of measurements which could be made on the objects to be classified. It is therefore necessary to choose a finite subset of these which leads to good classification results. However, since it has proved unreliable to rely on intuition to select these variables, more formal statistical approaches have been adopted. Most of these methods begin with a large number of variables measured on the design set and seek to find an effective subspace of the complete set of variables.

It is pertinent to enquire why it is necessary to go to the trouble of selecting a smaller set. Why not simply use the original large set? There are several reasons for this, one of the most straightforward being cost. If it is excessively expensive or time-consuming to gather measurements, then the fewer the better. If an adequate subset of the original measurements can be found then only this subset need be measured on all future objects to be classified.

Another reason demonstrates the rather artificial nature of the division into variable selection and decision surface estimation. It may be that judicious transformation and selection will lead to a variable set permitting a simple yet effective decision surface. For example, the two-dimensional data set of Figure 6.1 needs a quadratic decision surface in the original (x, y) space, but only a linear decision surface in the space spanned by transformed variables $(x - a)^2$ and $(y - b)^2$. (If we make the further transformation to $r = (x - a)^2 + (y - b)^2$ the decision surface becomes trivial.) Of course, these two conceptualizations are merely different ways of looking at the same mathematical function, and for the small examples used to illustrate it the distinction between a linear function of transformed variables and a nonlinear function of the original variables is rather unreal. In practice, though, when we may be dealing with hundreds of variables, considerable advantages are obtained by splitting the mathematical formalism into two parts.

A third reason for reducing the dimensionality of the space in which classifications are made is simply to eliminate redundancy. There is no point in

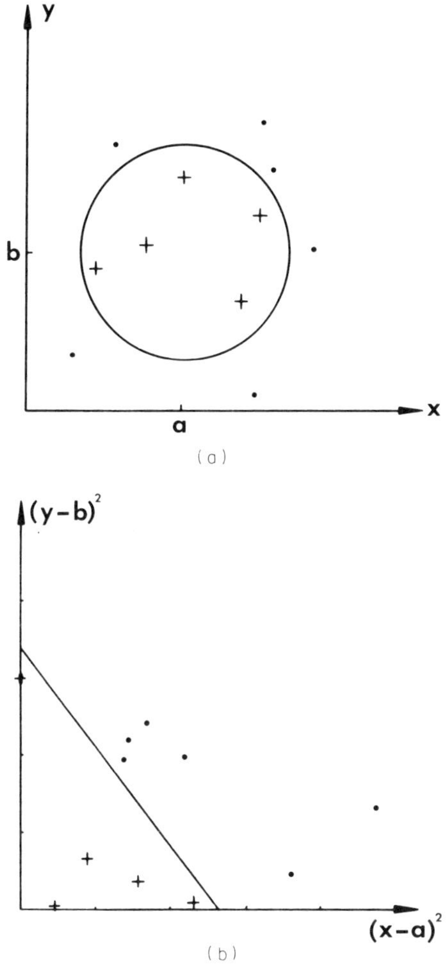

Figure 6.1 In the (x, y)-space (Figure (a)) a quadratic decision surface is needed, but in the transformed $(x - a)^2$, $(y - b)^2$-space (Figure (b)) a linear decision surface suffices

measuring a variable which does not add to the accuracy of the classification achieved without this variable.

Finally a very important reason for reducing the number of variables is discussed at length in Section 6.2. This is that a lower misclassification rate can sometimes be achieved by using fewer variables. Early workers in pattern recognition observed the superficially puzzling phenomenon that for a given data set as the number of measurements increased so the misclassification rate at first declined, but then began to increase. One might have expected that since each extra variable can only add information, and never subtract it,

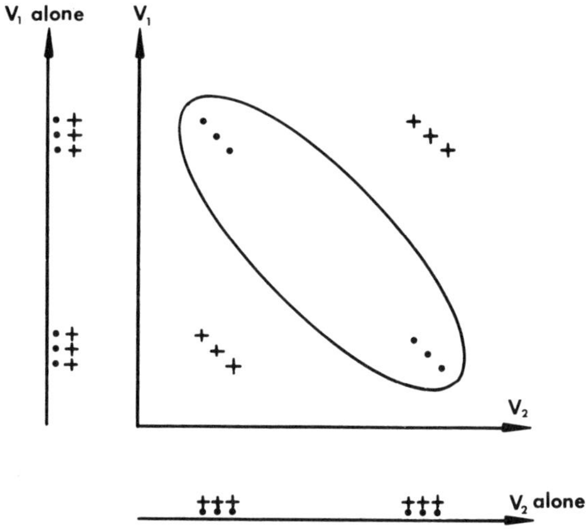

Figure 6.2 The importance of using multivariate relationships

the more variables the better. Several explanations for this effect have been proposed and these are outlined in Section 6.2.

During the process of selecting the final variables an indication can be obtained of the relative importance of individual variables and of combinations of variables. The latter is important. It is often the case that two individual variables are, by themselves, not very good discriminators. Taken in conjunction, however, they may be highly effective. Figure 6.2 illustrates two variables which, taken separately, are useless for discriminating between the two classes. When taken together, however, a simply quadratic decision surface perfectly separates the classes. This does not simply apply to two variables, of course, and in general it is *multivariate* relationships which are important, not simply univariate ones. And it is precisely because of the complexity of multivariate relationships that we need computer aid: human visualization abilities cannot satisfactorily handle such relationships.

6.2 DIMENSIONALITY AND MISCLASSIFICATION RATE

We commented above that early workers in pattern recognition observed the peculiar effect that, for a given design set, as the number of variables is increased (i.e. as more and more measurements are taken on the design set and the objects to be classified) so the classification performance of the resulting decision surface initially improved but then began to deteriorate. Several explanations for this phenomenon have been proposed, many of them based on the qualification that the design set be given—or, more accurately, that it be of a fixed finite size. An informal explanation along these lines might run as follows.

As the number of variables, d, increases so the number of parameters defining the decision surface increases, and this more flexible decision surface classifies the design set with a lower misclassification rate. However, as d increases so more and more of the probability associated with each class lies in regions of low pdf (a consequence of the curse of dimensionality). This means that the design set becomes more and more sparsely distributed, and the design set elements become less and less representative of the shape of the class conditional pdfs. The consequence is that the decision surface may fit the design set better with increasing d, but that this decision surface generalizes less well to new samples, i.e. the true error rate increases.

To introduce the idea from a classical statistical point of view let us initially ignore the error rate and concentrate on a more widely used statistic, namely Hotelling's T^2; that is, the distance between the two sample means relative to the dispersion within the samples. Formally

$$T^2 = \frac{n_1 n_2}{n}(\bar{\mathbf{x}}_1 - \bar{\mathbf{x}}_2)' S^{-1}(\bar{\mathbf{x}}_1 - \bar{\mathbf{x}}_2)$$

$$= \frac{n_1 n_2}{n} D^2$$

where $\bar{\mathbf{x}}_i$ is the mean for class ω_i, S is the assumed common variance–covariance matrix, and D^2 is the squared Mahalanobis distance (this latter appears elsewhere in this book). To investigate whether the samples come from different populations, one asks how often one would expect to observe a T^2 as large or larger than the T^2 estimated from the samples if the two populations are identical. (And if the probability of such a large T^2 is sufficiently low one tentatively—with a certain risk of error—concludes that the populations are distinct.) Formally, the value of

$$J = \frac{n - 1 - d}{(n - 2)d} T^2$$

is compared with the F distribution with d and $(n - 1 - d)$ degrees of freedom.

Our aim is to study how the distance between the two samples, as measured by T^2, changes as d increases. Thus, for simplicity let us suppose that each variable is independent of every other variable and that the standardized difference between the sample means is some constant k for each variable. This is something of an ideal situation in that normally the variables will be correlated—we can thus expect real data to exhibit behaviour worse than that illustrated below. Our assumptions mean that $D^2 = k^2 d$ and therefore that

$$J = (n - 1 - d)n_1 n_2 k^2 d/(n - 2)dn$$

$$= \frac{(n - 1)n_1 n_2 k^2}{(n - 2)n} - d\frac{n_1 n_2 k^2}{(n - 2)n}$$

This is a linear function of d, decreasing as d increases.

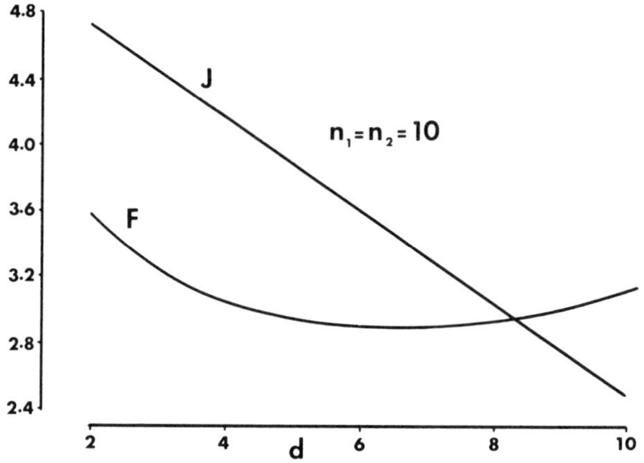

Figure 6.3 Plot of the 5 per cent F-value and matching J-value for two samples each of size 10 as the number of dimensions, d, increases

An example of a J function (with the above assumptions about the variables) and matching 5 per cent level values of F is given in Figure 6.3. Recall that the F/J relationship is interpreted as rejecting the null hypothesis that the samples are from the same class if $J > F$. It is obvious that as d increases so the probability of rejecting this hypothesis at first increases and then decreases. And this is despite our assumption that each variable separates the two samples. Figure 6.4(a) and (b) illustrates $(J - F)$ for several values of n. It is clear that, for larger n, d has to be proportionately larger before the deterioration in performance becomes apparent.

Liddell (1977) provides some examples of how increasing d has led to decreasing J.

Returning now to the error rate itself, we find that for normal classes a number of authors have studied the rate at which D^2 must increase as d increases in order to maintain the error rate constant or decrease it. Thus, Van Ness and Simpson (1976) and Van Ness (1979) apply Monte Carlo methods to compare five discriminant analysis algorithms: linear discriminant analysis with known variance–covariance matrix, linear discriminant analysis with unknown variance–covariance matrix, quadratic discriminant analysis with unknown variance–covariance matrices, and two kernel methods, one with normal kernels and the other with Cauchy kernels. They drew samples from two multivariate normal distributions with parameters

$$\boldsymbol{\mu}_1' = (0, \ldots, 0); \qquad \boldsymbol{\mu}_2' = (\delta, 0, \ldots, 0)$$
$$\boldsymbol{\Sigma}_1 = \boldsymbol{\Sigma}_2 = \mathbf{I},$$

and equal priors. For these populations $\Delta^2 = \delta^2$ is the squared Mahalanobis distance—and this remains constant as d increases. For each method they then

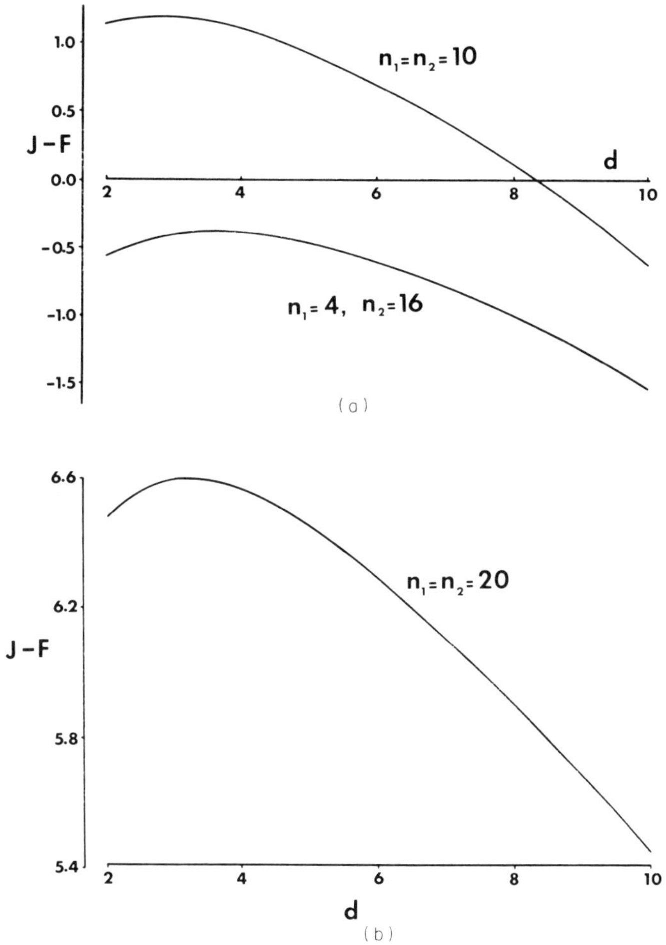

Figure 6.4 (a) Plots of $J - F$ (5 per cent) for $n_1 = 10$, $n_2 = 10$, and $n_1 = 4$, $n_2 = 16$ as d increases. (b) Plot of $J - F$ (5 per cent) for $n_1 = n_2 = 20$ as d increases

give two types of plot:

(i) P (correct classification) against δ for $d = 1, 2, 3, 5, 10, 20$, and 30.
(ii) δ against d for P (correct classification) fixed at 0.6, 0.7, and 0.8.

The first type of plot can be used to determine the discriminatory power lost by increasing d with δ fixed. The second type shows how much δ must increase in order to justify increasing d. It is particularly interesting to note that even though the data was from normal populations Van Ness and Simpson found 'the nonparametric algorithms to be very stable with increasing dimension and to outperform the linear and quadratic (unknown covariances) algorithms at much smaller dimensions than expected'.

Van Ness (1979) has extended this work to unequal variance–covariance matrices. Specifically, he produces the same kinds of plots when $\mathbf{\Sigma}_1 = \mathbf{I}$ and $\mathbf{\Sigma}_2 = \mathbf{I}/2$. The curves can be used in the same way as before. The algorithms he compares here are:

(a) quadratic discriminant analysis with known variance–covariance matrices;

(b) linear discriminant analysis with unknown variance–covariance matrices;

(c) quadratic discriminant analysis with unknown variance–covariance matrices;

(d) the kernel method using normal kernels and the same spread parameters for each population;

(e) the kernel method using normal kernels and permitting different spread parameters in each of the two populations; and

(f) an average link method which classifies a point $\mathbf{x}$ into that class whose design set elements have the smallest coverage distance from $\mathbf{x}$.

(Once again the general conclusion was that the non-parametric algorithms were stable at high dimensionality. Van Ness also observed a marked difference in performance between (d) and (e) above, the latter being better.)

Jain and Waller (1978) and El-Sheikh and Wacker (1980) used both numerical and analytic methods to study the relationship between d, Δ^2, and error rate for the special case of two multivariate normal classes with equal priors. Jain and Waller used the classical Fisher linear discriminant function, as outlined in Chapter 4, and demonstrated that if $\Delta^2 \gg 4d/n$ then the error rate will not increase when d is increased to $(d + 1)$ if

$$\Delta^2_{d+1} - \Delta^2_d = \Delta^2_d/(n - 3 - d)$$

They also give optimal values for d in two special cases. El-Sheikh and Wacker assume the within-class variance–covariance matrix to be known and show that if Δ^2_d increases faster than $\sqrt{d}$ the error rate will decrease monotonically with d, to zero in the limit.

The discussion so far has concentrated on one type of cause for the 'peaking' of classification accuracy (i.e. the way the accuracy at first increases and then decreases as d increases). Other phenomena also play a part. To see this, first consider a theorem due to Waller and Jain (1978) (Van Campenhout, 1978, also proves a version of this theorem):

Theorem

The performance of a Bayesian classifier is not degraded by using a second input set if the first input set is recoverable from the second. That is to say, for a Bayesian classifier

$$\int P(\text{correct}\,|\,\mathbf{x}, \mathbf{x}_1, \ldots, \mathbf{x}_n; \boldsymbol{\theta})p(\mathbf{x}, \mathbf{x}_1, \ldots, \mathbf{x}_n\,|\,\boldsymbol{\theta})f(\boldsymbol{\theta})\, \mathrm{d}\mathbf{x}\, \mathbf{dx}_1 \ldots \mathrm{d}\mathbf{x}_n$$

$$\geqslant \int P(\text{correct}\,|\,\mathbf{y}, \mathbf{y}_1, \ldots, \mathbf{y}_n; \boldsymbol{\theta})p(\mathbf{y}, \mathbf{y}_1, \ldots, \mathbf{y}_n\,|\,\boldsymbol{\theta})f(\boldsymbol{\theta})\, \mathrm{d}\mathbf{y}\, \mathrm{d}\mathbf{y}_1 \ldots \mathrm{d}\mathbf{y}_n,$$

for all pdfs $f(\boldsymbol{\theta})$ if there is a known function G such that $G(\mathbf{x}, \mathbf{x}_1, \ldots, \mathbf{x}_n) = (\mathbf{y}, \mathbf{y}_1, \ldots, \mathbf{y}_n)$. ■

Note that Bayesian here means 'predictive' in the sense of Chapter 3; that is, the pdf estimate is a Bayesian estimate obtained as a weighted combination of densities (see Section 3.3). This is in contrast to the methods discussed so far in this section where pdf estimates are obtained by substituting estimates for parameters.

The implication of this theorem is that for Bayesian classifiers one should not expect the error rate to fall and then begin to increase again as d (or, more generally, the measurement complexity—see below) increases. However, consider the results of Hughes (1968). He investigated the case of discrete variables, using as the performance criterion the mean correct recognition probability, where the mean is taken over all probability stuctures with given sample size (n), priors (p_1 and p_2), and measurement complexity $G = \Pi_{i=1}^{d} g_i$ (where g_i is the number of categories in variable i). This means that no distribution assumptions at all are being made.

Hughes shows that the infinitive data set recognition accuracy is given by

$$A(G, p_1) = p_1 + p_2(G-1)\left(\frac{p_1}{p_2}\right)^G \sum_{j=0}^{G} \frac{G!}{j!\,(G-j)!\,(2G-j-1)[p_1/(1-2p_1)]^j}$$

(where p_1 is assumed less than p_2). When $p_1 = p_2$ we have

$$A(G, \tfrac{1}{2}) = (3G-2)/(4G-2)$$

Points to observe are that A increases monotonically from $\max(p_1, p_2)$ and converges asymptotically to $(1 - p_1 p_2)$, which can be considerably less than 100 per cent if p_1 is not near 0 or 1.

For finite data sets the use of the *mean* recognition accuracy (an implicit adoption of Bayesian methods) means that Carlo Monte methods can be avoided—though it also means that a particular structure could have a better or worse performance than the average result obtained. Using the classification rule (with $p_1 \leq p_2$)

$$\mathbf{x} \in \begin{cases} \Omega_1, & \text{if } n_{1\mathbf{x}} > n_{2\mathbf{x}} \\ \Omega_2, & \text{otherwise} \end{cases}$$

(with $n_{i\mathbf{x}}$ the number of observations from the class i design set which fall in cell $\mathbf{x}$) Hughes obtained:

$$A(G, n, p_1) = \sum_{r=0}^{n2} \sum_{s=0}^{n1} \left[\frac{(n_2-r+1)(n_2-r+2)\ldots(n_2-r+G-2)}{(n_2+1)(n_2+2)\ldots(n_2+G-1)}\right] \times \left[\frac{(n_1-s+1)(n_1-s+2)\ldots(n_1-s+G-2)}{(n_1+1)(n_1+2)\ldots(n_1+G-1)}\right] g(r, s)$$

with

$$g(r, s) = \begin{cases} G(G-1)^2 p_1 \dfrac{s+1}{n_1+G}, & \text{if } s > r \\ G(G-1)^2 p_2 \dfrac{r+1}{n_2+G}, & \text{if } s \leqslant r \end{cases}$$

Curves illustrating this function are given in Figure 6.5. They demonstrate clearly an initial improvement and later deterioration in performance. This seems to contradict the above theorem.

In fact, as both Waller and Jain (1978) and Van Campenhout (1978) show, this is due to the models which Hughes uses at different levels of measurement complexity being non-comparable. (Starting from a given prior distribution $f(\theta)$ at $d = d_1$, reducing d imposes restrictions on the class of priors at $d = d_2 < d_1$, restrictions which Hughes's approach violates. Details are given in the references.)

Hughes also notes that if one class is more likely than the other then if G/n is large enough the above classification rule is inferior to classifying purely on the basis of the prior probabilities. He states: 'If insufficient sample data are available to estimate the pattern probabilities accurately, then a Bayes recogniser is not necessarily optimal'. Abend and Harley(1969), however, point out that in fact Hughes's rule is not the Bayes rule (meaning minimum average risk) but rather uses the Bayes rule that would be appropriate if the cell probabilities were known and then substitutes estimates for them since they

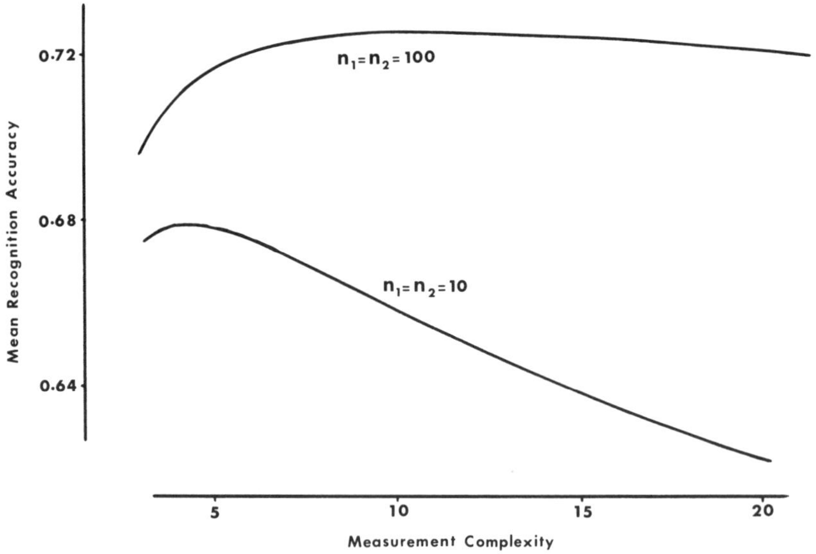

Figure 6.5 The mean recognition accuracy for samples of size $n_1 = n_2 = 100$ and $n_1 = n_2 = 10$ plotted against the measurement complexity G

are in fact unknown. Chandrasekaran and Harley (1969) show that the true Bayes rule (with, again, $p_1 \leq p_2$) given by

$$\mathbf{x} \in \begin{cases} \Omega_1, & \text{if } p_1 \dfrac{n_{1\mathbf{x}} + 1}{n_1 + G} > p_2 \dfrac{n_{2\mathbf{x}} + 1}{n_2 + G} \\ \Omega_2, & \text{otherwise} \end{cases}$$

is in fact never worse than choosing the most probable class.

Murray (1977) has investigated a slightly different aspect of the peculiar relationship between error rate and dimensionality for continuous variables. Instead of using the design set to estimate the decision surface parameters he fixes these at their optimum values (known since his data is artificial) and uses the design set to select d' from d variables. When d' is near $d/2$ there are a relatively large number of possible subsets $\left(\text{approximately} \binom{d}{d/2}\right)$ so that there is considerable freedom to choose a highly discriminating subset. However, when d' is near to d (or 0) there is little freedom, and since the parameters of the decision surface are preset at their optimum values the decision surface is almost completely determined (and is completely determined when $d = d'$). Thus, as d' moves from 0 to $d/2$ the apparent error rate decreases because of the freedom to choose the variables whose associated surface matches the data. But as d' moves from $d/2$ to d the apparent error rate increases, again yielding a U-shaped curve.

In fact, of course, the decision surface parameters must also be estimated from the design set.

Further light may be shed on the relation between n, d, and the error rate by considering an aspect of linear decision surfaces—in particular let us consider *linear dichotomies*. A dichotomy of a set of points is an allocation of each of them to one of two mutually exclusive classes. A linear dichotomy is such an allocation in which the two classes can be perfectly separated by a linear surface. For convenience we shall make the further assumption that no subset of $(d + 1)$ points lie in a $(d - 1)$-dimensional subspace. Letting the total number of dichotomies of n points in d dimensions be $T(n, d)$ and the number of these which are linear be $L(n, d)$, we have

$$\frac{L(n, d)}{T(n, d)} = \begin{cases} 1, & n < d \\ \dfrac{1}{2^{n-1}} \displaystyle\sum_{i=0}^{d} \binom{n-1}{i}, & n \geq d \end{cases}$$

Figure 6.6 gives examples of this function. Note that for $n = 2(d + 1)$ the probability is $\frac{1}{2}$ that any arbitrary allocation will be perfectly separable by a linear surface. The conclusion is that to avoid spurious results which do not generalize, it is necessary to have n several times as large as d.

In this section we have discussed the relationship between dimensionality and error rate and may have given the impression that problems and (superfi-

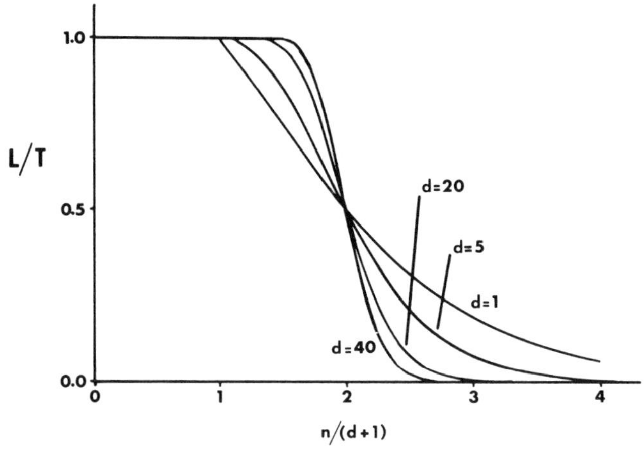

Figure 6.6 The proportion of dichotomies which are linear plotted against the ratio $n/(d + 1)$

cial) paradoxes cannot arise for small d. However, this is not necessarily the case and a cautionary note is in order. Many methods do not minimize the error rate (on the design set) directly but instead make some simplifying assumptions and then try to minimize the error rate of the model based on these assumptions. For example, the classical linear discriminant function method outlined in Section 4.5 assumes that the class-conditional pdfs can be described adequately by only the first- and second-order moments and then tries to maximize the ratio

(between-class distance)/(within-class dispersion)

If the class-conditional pdfs are normal with equal variance–covariance matrices, then such an approach is equivalent to minimizing the error rate on the design set. If the pdfs are not normal with such variance–covariance matrices, however, then the two approaches need not be equivalent, though the hope is they they will still be closely related. To illustrate, consider the data shown in Figure 6.7. This shows the design set of 15 dyslexic and 18 control subjects from a study of reading and writing difficulty. (I am indebted to Professor Gerald Russell for permission to include this data set.) They have been measured on two variables: a phonetic reading test and the Schonell writing test. Line A shows the decision surface resulting when Fisher's linear discriminant function method is applied using just the writing test (horizontal axis). As can be seen, one of the dyslexic subjects is misclassified even though the design set error rate could be reduced by moving the decision surface to the right. When we consider the two variables together the same sort of phenomenon occurs (line B). Only now, maximal separability between the two classes, as measured by the above ratio, is obtained when two points are misclassified. Once again, a classifier which set out to minimize the error rate

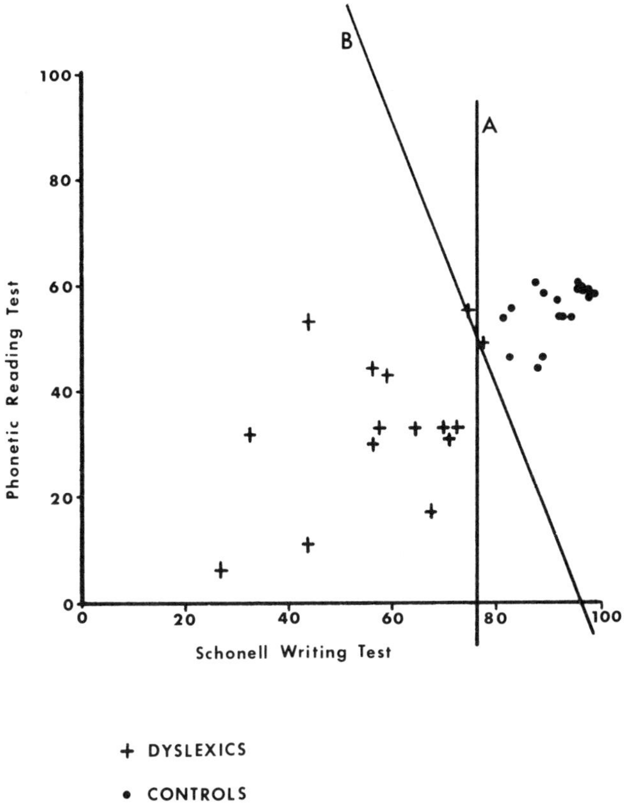

Figure 6.7 Design set of 15 dyslexics and 18 controls from a study of reading and writing difficulty

on the design set could correctly classify these two points. If the model was correct (i.e. normal class conditional pdfs with equal variance–covariance matrices) then error rate would be inversely monotonically related to the above ratio. However, it is obvious from Figure 6.7 that the classes have different variances. (Note that, even if the population distributions satisfy the model, the distribution of a small sample could deviate significantly from it.)

Finally, notice that in this example as d increases from 1 (writing test alone) to 2 (reading and writing test) the design set misclassification rate increases from 3 to 6.1 per cent so that increasing dimensionality appears to increase the error rate.

6.3 CLASS SEPARABILITY MEASURES

In Sections 6.1 and 6.2 we discussed some of the reasons for finding a subspace of the original d-dimensional space in which to carry out the classification. Clearly the error rate associated with each subspace would be the best way of choosing between subspaces. Unfortunately, there are practical problems (outlined below) associated with this straightforward approach so other measures of how well we can discriminate between classes must be used. For obvious reasons those measures are termed *separability* measures.

The problem of cluster analysis, considered in Chapter 7, is in many ways similar to that of variable selection. The difference is that here we choose a subspace to optimize separability and there we choose the *class of each element* to optimize separability. Note, however, that it may be necessary to select a subspace prior to a cluster analysis. If such is the case the aim is to choose a subspace which retains as much of the *structure* of the data as possible. Here only that aspect of the structure which is reflected in the separability measure is important.

For the moment we return to the ideal solution: consider each eligible subspace (how to find these is discussed in the next two sections), compute the best classifier for each eligible subspace, and from this estimate the error rate. That subspace which yields the smallest error rate is adopted.

From this ideal approach we can see at least one problem which will arise in general, i.e. that since the error rate depends on the classifier used, the best set of variables will also depend on the classifier.

A second difficulty which limits the applicability of the ideal approach is that when selecting a subset of variables (either without or after a transformation) a very large number of alternatives exist, each of which must be examined (at least implicitly—see Section 6.4.2) in order to find the best. Thus, considerable advantages result from using a separability measure which is quick to evaluate. Explicit evaluation of the true error rate of a given classifier may well not be quick enough.

These difficulties mean that certain simplifying assumptions have necessarily to be adopted during the variable selection process. For example, it is common to evaluate subspaces on the basis of one classifier and then apply another one to actually classify new observations. This is particularly true of those methods which utilize transformations.

Despite the problems it might very occasionally be possible to apply direct non-parametric estimation of the Bayes error rate. We shall defer discussion of such methods to Chapter 8, contenting ourselves here with the remark that the methods of Chapter 2 can be applied. In effect such approaches can be viewed as Monte Carlo methods of performing the multivariate integration in (for the two-class case)

$$\varepsilon = \int_{\Omega_1} p(\omega_2)p(\mathbf{x}\,|\,\omega_2)\,\mathrm{d}\mathbf{x} + \int_{\Omega_2} p(\omega_1)p(\mathbf{x}\,|\,\omega_1)\,\mathrm{d}\mathbf{x}$$

Patrick and Fisher (1969) develop such a non-parametric approach and base a transformation method (see Section 6.5) on it. They conclude, however: 'The computation time thus limits the usefulness of the nonparametric approach.' In an attempt to get around this they suggest using a subsample of the complete design set during the initial steps of their (iterative) estimation of a good subspace. This idea can be applied perfectly generally, of course.

A quicker evaluation would be possible if we could express ε in a closed form, so allowing us to sidestep the multivariate integration. If the two populations have normal distributions with identical covariance matrices ($\mathbf{\Sigma}_1 = \mathbf{\Sigma}_2 = \mathbf{\Sigma}$) we have, as we showed in Chapter 4 a linear decision surface

$$\mathbf{x}'\mathbf{\Sigma}^{-1}(\boldsymbol{\mu}_1 - \boldsymbol{\mu}_2) - \tfrac{1}{2}(\boldsymbol{\mu}_1'\mathbf{\Sigma}^{-1}\boldsymbol{\mu}_1 - \boldsymbol{\mu}_2'\mathbf{\Sigma}^{-1}\boldsymbol{\mu}_2) \gtrless \ln \frac{P(\omega_2)}{P(\omega_1)} \Rightarrow \mathbf{x} \in \begin{cases} \Omega_1 \\ \Omega_2 \end{cases}$$

Letting $f(\mathbf{x})$ represent the left-hand side of the expression we see that $f(\mathbf{x})$ is a linear function of $\mathbf{x}$. Thus, since for each class we are assuming the $\mathbf{x}_i$ to be normally distributed, for class ω_i the mean of $f(\mathbf{x})$ is

$$\begin{aligned} v_i &= \boldsymbol{\mu}_i'\mathbf{\Sigma}^{-1}(\boldsymbol{\mu}_1 - \boldsymbol{\mu}_2) - \tfrac{1}{2}[\boldsymbol{\mu}_1'\mathbf{\Sigma}^{-1}\boldsymbol{\mu}_1 - \boldsymbol{\mu}_2'\mathbf{\Sigma}^{-1}\boldsymbol{\mu}_2] \\ &= \left(\boldsymbol{\mu}_i - \frac{\boldsymbol{\mu}_1}{2} - \frac{\boldsymbol{\mu}_2}{2}\right)'\mathbf{\Sigma}^{-1}(\boldsymbol{\mu}_1 - \boldsymbol{\mu}_2) \\ &= (-1)^{i+1}\,\tfrac{1}{2}(\boldsymbol{\mu}_1 - \boldsymbol{\mu}_2)'\mathbf{\Sigma}^{-1}(\boldsymbol{\mu}_1 - \boldsymbol{\mu}_2) = \tfrac{1}{2}\Delta^2(-1)^{i+1} \end{aligned}$$

(where Δ^2 is the population version of the D^2 referred to above and to be discussed later) and the variance is

$$\sigma_i^2 = (\boldsymbol{\mu}_1 - \boldsymbol{\mu}_2)'\mathbf{\Sigma}^{-1}(\boldsymbol{\mu}_1 - \boldsymbol{\mu}_2) = \Delta^2$$

The overall probability of error is thus

$$\tfrac{1}{2}P[f(\mathbf{x}) > \ln P(\omega_2)/P(\omega_1)\,|\,\omega_2] + \tfrac{1}{2}P[f(\mathbf{x}) < \ln P(\omega_2)/P(\omega_1)\,|\,\omega_1]$$

$$= \frac{1}{2}\int_t^{\infty} \Phi(v_2, \sigma_2^2) + \frac{1}{2}\int_{-\infty}^{t} \Phi(v_1, \sigma_1^2)$$

(where $t = \ln P(\omega_2)/P(\omega_1)$ and $\Phi(a, b^2)$ is the normal distribution with mean a and variance b^2). This is not a closed form but sample estimates could be substituted for v_i and σ_i^2 and the result obtained from normal tables.

The considerable simplification and hence quicker evaluation which results if a closed form can be found for the separability criterion has led to the development of other criteria which have closed form solutions for certain pdfs and which, while not being directly expressible in terms of the Bayes error, are related to it. Desirable attributes of such measures are:

(a) $J(\omega_1, \omega_2) = 0$, if $p(\mathbf{x}|\omega_1) = p(\mathbf{x}|\omega_2)$,
(b) $J(\omega_1, \omega_2) \geqslant 0$,
(c) $J(\omega_1, \omega_2)$ attains its maximum if the classes are disjoint.

An example of a measure satisfying conditions (a) to (c) is the *divergence*

$$J_1(\omega_1, \omega_2) = \int [p(\mathbf{x}|\omega_1) - p(\mathbf{x}|\omega_2)] \ln \frac{p(\mathbf{x}|\omega_1)}{p(\mathbf{x}|\omega_2)}\, d\mathbf{x}$$

For normal classes with unequal covariance matrices $\mathbf{\Sigma}_1$ and $\mathbf{\Sigma}_2$ this becomes

$$\tfrac{1}{2}(\boldsymbol{\mu}_1 - \boldsymbol{\mu}_2)'(\boldsymbol{\Sigma}_1^{-1} + \boldsymbol{\Sigma}_2^{-1})(\boldsymbol{\mu}_1 - \boldsymbol{\mu}_2) + \tfrac{1}{2}\,\mathrm{tr}(\boldsymbol{\Sigma}_1^{-1}\boldsymbol{\Sigma}_2 + \boldsymbol{\Sigma}_2^{-1}\boldsymbol{\Sigma}_1 - 2\mathbf{I})$$

which reduces to Mahalanobis's Δ^2

$$(\boldsymbol{\mu}_1 - \boldsymbol{\mu}_2)'\boldsymbol{\Sigma}^{-1}(\boldsymbol{\mu}_1 - \boldsymbol{\mu}_2)$$

when $\boldsymbol{\Sigma}_1 = \boldsymbol{\Sigma}_2$. Kullback (1968) deals with the divergence in some detail.

Another measure which has received attention is the *Chernoff distance*

$$J_2(\omega_1, \omega_2) = -\ln \int p(\mathbf{x}\,|\,\omega_1)^{1-s} p(\mathbf{x}\,|\,\omega_2)^s \, d\mathbf{x}$$

where $s \in [0, 1]$. This measure is particularly interesting because it provides an upper bound for the Bayes error ε

$$\varepsilon \leqslant P(\omega_1)^{1-s} P(\omega_2)^s \exp(-J_2)$$

for $s \in [0, 1]$.

The special case with $s = \tfrac{1}{2}$

$$J_3(\omega_1, \omega_2) = -\ln \int \sqrt{p(\mathbf{x}\,|\,\omega_1)p(\mathbf{x}\,|\,\omega_2)}\, d\mathbf{x}$$

is known as the *Bhattacharyya distance* and has received widespread attention. This special case, as well as providing the upper bound common to all Chernoff distances, also provides a lower bound on ε so that

$$\tfrac{1}{2}[1 - (1 - 4P(\omega_1)P(\omega_2)\exp(-2J_3))^{1/2}] \leqslant \varepsilon \leqslant \sqrt{P(\omega_1)P(\omega_2)}\,\exp(-J_3)$$

For normal classes with $\boldsymbol{\Sigma}_1 \neq \boldsymbol{\Sigma}_2$ the general Chernoff measure becomes

$$\tfrac{1}{2}s(1 - s)(\boldsymbol{\mu}_1 - \boldsymbol{\mu}_2)'\{(1 - s)\boldsymbol{\Sigma}_1 + s\boldsymbol{\Sigma}_2\}^{-1}(\boldsymbol{\mu}_1 - \boldsymbol{\mu}_2) + \tfrac{1}{2}\ln \frac{|(1 - s)\boldsymbol{\Sigma}_1 + s\boldsymbol{\Sigma}_2|}{|\boldsymbol{\Sigma}_1|^{1-s}|\boldsymbol{\Sigma}_2|^s}$$

and the particular Bhattacharyya measure becomes

$$\tfrac{1}{8}(\boldsymbol{\mu}_1 - \boldsymbol{\mu}_2)'\left(\frac{\boldsymbol{\Sigma}_1 + \boldsymbol{\Sigma}_2}{2}\right)^{-1}(\boldsymbol{\mu}_1 - \boldsymbol{\mu}_2) + \tfrac{1}{2}\ln \frac{|\tfrac{1}{2}(\boldsymbol{\Sigma}_1 + \boldsymbol{\Sigma}_2)|}{|\boldsymbol{\Sigma}_1|^{1/2}|\boldsymbol{\Sigma}_2|^{1/2}}$$

(with $\boldsymbol{\Sigma}_1 = \boldsymbol{\Sigma}_2$ both the Chernoff and Bhattacharyya measures become proportional to Mahalanobis's Δ^2).

To illustrate, 100 random points were generated from a mixture of two bivariate normal distributions with parameters

$$P(\omega_1) = 0.5; \quad \begin{pmatrix}\mu_{11}\\ \mu_{12}\end{pmatrix} = \begin{pmatrix}0\\ 0\end{pmatrix}; \quad \boldsymbol{\Sigma}_1 = \begin{pmatrix}1 & 0\\ 0 & 2\end{pmatrix}$$

$$P(\omega_2) = 0.5; \quad \begin{pmatrix}\mu_{21}\\ \mu_{22}\end{pmatrix} = \begin{pmatrix}2\\ 0\end{pmatrix}; \quad \boldsymbol{\Sigma}_2 = \begin{pmatrix}2 & 0\\ 0 & 1\end{pmatrix}$$

The maximum likelihood estimates of these parameters were

$$\hat{P}(\omega_1) = 0.46; \quad \begin{pmatrix}\hat{\mu}_{11}\\ \hat{\mu}_{12}\end{pmatrix} = \begin{pmatrix}0.213\\ -0.608\end{pmatrix}; \quad \hat{\boldsymbol{\Sigma}}_1 = \begin{pmatrix}1.395 & 0.448\\ 0.448 & 3.107\end{pmatrix}$$

$$\hat{P}(\omega_2) = 0.54; \qquad \begin{pmatrix} \hat{\mu}_{21} \\ \hat{\mu}_{22} \end{pmatrix} = \begin{pmatrix} 2.166 \\ 0.109 \end{pmatrix}; \qquad \hat{\mathbf{\Sigma}}_2 = \begin{pmatrix} 4.420 & -0.022 \\ -0.022 & 0.904 \end{pmatrix}$$

From these the Bhattacharyya measure gives lower bound 0.14 and upper bound 0.35 on the Bayes error. The latter is, in fact, 0.21.

Kailath (1967) also gives some closed forms for the Bhattacharyya distance under various distributions. For example:

Multinomial

$$J_3 = -\ln\left(\sum_{\mathbf{x}} \sqrt{p(\mathbf{x}\,|\,\omega_1)p(\mathbf{x}\,|\,\omega_2)}\right)$$

where the summation is over all cells.

Poisson

$$J_3 = \tfrac{1}{2}(\sqrt{\lambda_1} - \sqrt{\lambda_2})^2$$

where λ_1 is the parameter for population i (so the pdf of population i is $p_i(n) = e^{-\lambda_i}\lambda_i^n/n!$).

If one is uneasy about assuming particular distributions for the class-conditional pdfs then the above measures hold no advantage over the basic Bayes error rate—each still requires a multivariate integration. Acknowledging this fact, other measures have been developed from intuitive ideas based on the scatter matrices. Their advantage of easy evaluation is coupled with the disadvantage that they have no general relationship with the Bayes error rate. Since they are based solely on the scatter matrices, which are second order, we would expect the resulting measures to be closely related to those discussed above under the assumption of normal populations. This indeed is the case. Another point to note is the fundamental nature of Mahalanobis's Δ^2 which, as we have seen above, is simply the Euclidean distance between $\boldsymbol{\mu}_1$ and $\boldsymbol{\mu}_2$ after the space has been normalized by $\mathbf{\Sigma}^{-1}$. (The same comments apply to Hotelling's T^2 which was originally chosen as the multivariate extension of t^2 because of its invariance under linear transformations.) So, expecting the fundamental Δ^2 to show up somewhere, let us consider some of the measures based on the scatter matrices. We begin by defining the *total* scatter matrix of a set of data points as (with two classes)

$$\mathbf{T} = \sum_{j=1}^{2} \sum_{i=1}^{n_j} (\mathbf{x}_{ij} - \bar{\mathbf{x}})(\mathbf{x}_{ij} - \bar{\mathbf{x}})'$$

where $\mathbf{x}_{ij}$ is the ith point from class j, n_j is the total number of sample points in class j, and $\bar{\mathbf{x}}$ is the overall sample mean.

The *within*-class scatter matrix is defined as

$$\mathbf{W} = \sum_{j=1}^{2} \sum_{i=1}^{n_j} (\mathbf{x}_{ij} - \bar{\mathbf{x}}_j)(\mathbf{x}_{ij} - \bar{\mathbf{x}}_j)'$$

where $\bar{\mathbf{x}}_j$ is the sample mean for class j, and the *between*-class scatter matrix is

$$\mathbf{B} = \frac{n_1 n_2}{n_1 + n_2} (\bar{\mathbf{x}}_1 - \bar{\mathbf{x}}_2)(\bar{\mathbf{x}}_1 - \bar{\mathbf{x}}_2)'$$

If $\mathbf{B}$ is rewritten as

$$\mathbf{B} = \sum_{j=1}^{2} n_j (\bar{\mathbf{x}}_j - \bar{\mathbf{x}})(\bar{\mathbf{x}}_j - \bar{\mathbf{x}})' = \sum_{j=1}^{2} \sum_{i=1}^{n_j} (\bar{\mathbf{x}}_j - \bar{\mathbf{x}})(\bar{\mathbf{x}}_j - \bar{\mathbf{x}})'$$

a little algebra reveals that $\mathbf{T} = \mathbf{W} + \mathbf{B}$.

Since our aim is to use $\mathbf{T}$, $\mathbf{W}$, and $\mathbf{B}$ as the basis for a separability measure we must somehow find a univariate function which can be optimized. Clearly we want, in some sense, to maximize $\mathbf{B}$ relative to $\mathbf{W}$, the between-class spread relative to the within-class spread. For example, we might try to find the subspace in which some univariate function of $\mathbf{W}^{-1}\mathbf{B}$ is maximized. A popular univariate criterion is tr $\mathbf{W}^{-1}\mathbf{B}$, the sum of the spreads of $\mathbf{W}^{-1}\mathbf{B}$ in the direction of the principal components of $\mathbf{W}^{-1}\mathbf{B}$. (This is derived more formally in Section 6.5.) Thus

$$J = \operatorname{tr} \mathbf{W}^{-1}\mathbf{B} = \frac{n_1 n_2}{n_1 + n_2} \operatorname{tr} \mathbf{W}^{-1}(\bar{\mathbf{x}}_1 - \bar{\mathbf{x}}_2)(\bar{\mathbf{x}}_1 - \bar{\mathbf{x}}_2)'$$

$$= \frac{n_1 n_2}{n_1 + n_2} (\bar{\mathbf{x}}_1 - \bar{\mathbf{x}}_2)'\mathbf{W}^{-1}(\bar{\mathbf{x}}_1 - \mathbf{x}_2)$$

$$= \frac{T^2}{n_1 + n_2} = \frac{n_1 n_2}{(n_1 + n_2)^2} D^2$$

Thus, this criterion (which, as we shall see, is that used in classical discriminant analysis) chooses between subspaces on the basis of the estimate D^2 of the Mahalanobis's Δ^2 between the two samples.

Another univariate function based on these matrices is $|\mathbf{W}|/|\mathbf{T}|$. Here the aim is to minimize $|\mathbf{W}|$ with respect to $|\mathbf{T}| = |\mathbf{W} + \mathbf{B}|$. We have

$$\frac{|\mathbf{W}|}{|\mathbf{T}|} = \frac{|\mathbf{W}|}{|\mathbf{W} + \mathbf{B}|} = \frac{|\mathbf{W}|}{|\mathbf{W}| \cdot |\mathbf{I} + \mathbf{W}^{-1}\mathbf{B}|} = \frac{1}{|\mathbf{I} + \mathbf{W}^{-1}\mathbf{B}|}$$

So, minimizing $|\mathbf{W}|/|\mathbf{T}|$ is equivalent to maximizing $|\mathbf{I} + \mathbf{W}^{-1}\mathbf{B}|$. Suppose now that λ is an eigenvalue of $\mathbf{W}^{-1}\mathbf{B}$, with $\mathbf{v}$ its corresponding eigenvector. Then

$$\mathbf{W}^{-1}\mathbf{B}\mathbf{v} - \lambda\mathbf{I}\mathbf{v} = 0$$

So $(\mathbf{W}^{-1}\mathbf{B} + \mathbf{I})\mathbf{v} - (\lambda + 1)\mathbf{I}\mathbf{v} = 0$. That is, if λ is an eigenvalue of $\mathbf{W}^{-1}\mathbf{B}$ then $(\lambda + 1)$ is an eigenvalue of $(\mathbf{I} + \mathbf{W}^{-1}\mathbf{B})$.

There are several ways in which these two-class ideas can be extended to the many-class case. A common approach is to maximize the separation between the two closest groups. This can be done using pairwise non-parametric estimation of between-class misclassification rates—and this has the drawback of the two-class case that it is very time-consuming—or any other measure of between-class separability can be used.

A different type of approach emerges when we note that the scatter matrices can easily be extended to the multi-class case. We have already noted that **B** can be rewritten as

$$\mathbf{B} = \sum_{j=1}^{2} \sum_{i=1}^{n_j} (\bar{\mathbf{x}}_j - \bar{\mathbf{x}})(\bar{\mathbf{x}}_j - \bar{\mathbf{x}})'$$

and it is then a simple matter to replace the summation over the two classes in **B**, **T**, and **W** by summations over N classes. Having done this we can then use the criteria tr $\mathbf{W}^{-1}\mathbf{B}$ and $|\mathbf{W}|/|\mathbf{T}|$ as before. The statistic $|\mathbf{W}|/|\mathbf{T}|$ (known as Wilks's Λ) is used in multivariate analysis of variance (see, for example, Kenward, 1979) where it is applied in precisely the manner we are using it to provide an overall measure of between-class differences.

Fukunaga and Short (1980) have discussed general criteria which depend only on first- and second-order moments, such as those derived from the scatter matrices shown, and which are invariant to non-singular linear transformations. We shall follow Fukunaga and Short and outline their results in terms of variance–covariance matrices rather than scatter matrices (the two sets being related by a constant factor) and we shall also, as they have done, use population rather than sample moments. Here $\boldsymbol{\mu}_i$ and $\boldsymbol{\Sigma}_i$ are the mean vector and variance–covariance matrix for class ω_i and $\boldsymbol{\mu}_0$ and $\boldsymbol{\Sigma}_0$ are the overall mean and variance–covariance matrix. Then for the two-class case Fukunaga and Short show that optimizing any criterion $f(\boldsymbol{\mu}_1, \boldsymbol{\mu}_2, \boldsymbol{\Sigma}_0)$ is equivalent to maximizing

$$Z^2 = (\boldsymbol{\mu}_1 - \boldsymbol{\mu}_2)'\boldsymbol{\Sigma}_0^{-1}(\boldsymbol{\mu}_1 - \boldsymbol{\mu}_2)$$

by choice of the new variables $\mathbf{y}(\mathbf{x}) = (y_1(\mathbf{x}), \ldots, y_{d'}(\mathbf{x}))'$. As an example, consider the criterion $J = \text{tr } \mathbf{W}^{-1}\mathbf{B}$ discussed above. In fact, as mentioned above, Fukunaga and Short consider $J = \text{tr } \mathbf{S}^{-1}\mathbf{U}$, where

$$\mathbf{S} = P(\omega_1)\boldsymbol{\Sigma}_1 + P(\omega_2)\boldsymbol{\Sigma}_2$$

and

$$\mathbf{U} = P(\omega_1)(\boldsymbol{\mu}_1 - \boldsymbol{\mu}_0)(\boldsymbol{\mu}_1 - \boldsymbol{\mu}_0)' + P(\omega_2)(\boldsymbol{\mu}_2 - \boldsymbol{\mu}_0)(\boldsymbol{\mu}_2 - \boldsymbol{\mu}_0)'$$

That this is of the form $f(\boldsymbol{\mu}_1, \boldsymbol{\mu}_2, \boldsymbol{\Sigma}_0)$ can be seen when it is rewritten as

$$J = \frac{2P(\omega_1)P(\omega_2)(\boldsymbol{\mu}_1 - \boldsymbol{\mu}_2)'\boldsymbol{\Sigma}_0^{-1}(\boldsymbol{\mu}_1 - \boldsymbol{\mu}_2)}{1 - P(\omega_1)P(\omega_2)(\boldsymbol{\mu}_1 - \boldsymbol{\mu}_2)\boldsymbol{\Sigma}_0^{-1}(\boldsymbol{\mu}_1 - \boldsymbol{\mu}_2)}$$

This is simply

$$J = \frac{2P(\omega_1)P(\omega_2)Z^2}{1 - P(\omega_1)P(\omega_2)Z^2}$$

so that maximizing J is indeed equivalent to maximizing Z^2.

For the $N(>2)$ class case Fukunaga and Short show that all criteria of the form $f(\boldsymbol{\mu}_1, \ldots, \boldsymbol{\mu}_N, \boldsymbol{\Sigma}_0)$ are equivalent to functions $g(Z_2, \ldots, Z_N)$ where

$$Z_i^2 = (\boldsymbol{\mu}_i - \boldsymbol{\mu}_1)'\boldsymbol{\Sigma}_0^{-1}(\boldsymbol{\mu}_i - \boldsymbol{\mu}_1)$$

i.e. functions of only the pairwise interclass distances.

More generally still, Fukunaga and Short consider criteria $f(\boldsymbol{\mu}_1, \ldots, \boldsymbol{\mu}_N, \boldsymbol{\Sigma}_1, \ldots. \boldsymbol{\Sigma}_N)$ and obtain a similar result, namely that all such criteria are weighted sums of the interclass distances using metrics induced by functions of the $\boldsymbol{\Sigma}_i$.

6.4 SELECTING THE VARIABLES

In this section we discuss algorithms for finding subspaces which are spanned by a subset of the original set of variables. We do not consider transformations which make implicit use of all the original variables: that will be considered in Section 6.5.

In principle the problem is a straightforward one: to find the best set of d' variables we merely consider each of the $\binom{d}{d'}$ possible sets, evaluate one of the criteria discussed in the preceding section for each of the sets, and choose the best set. In practice, however, this is often not feasible because $\binom{d}{d'}$ is very large. Not only does this mean that explicit error rate estimation cannot be used, but it also means that some other way must be found to accelerate the search. If the error rate was being used as a criterion to choose between variable sets one could identify the size of set and the particular set which minimized the criterion. However, since time limitations mean that we have to adopt one of the alternative criteria of Section 6.3 we have the further problem that these do not build up to a maximum and then decay as d increases (note that for these criteria a good set means a high value—the converse of the error rate). Thus, they cannot be used for a direct comparison between sets of different sizes, though it might be possible to use them in comparisons of sets of the same size.

One could avoid this problem by specifying beforehand the size of variable set one requires, or one might find the 'best' set of each size and compare these best using (say) an estimate of the true error rate. A more usual approach, however, is to use formal statistical tests to see if the improvement in discrimination afforded by the extra $(d - d')$ variables is statistically significant. Such tests are outlined in Sections 6.4.1, 6.4.3, and 6.5.

Returning to the d/error rate behaviour of the separability measures of Section 6.3, some of these measures (for example, those based on D^2) satisfy

$$J(X_1) \leqslant J(X_2) \leqslant \ldots \leqslant J(X_m)$$

for variable sets X_i such that

$$X_1 \subset X_2 \subset \ldots \subset X_m$$

This apparent disadvantage can be turned to our advantage as is demonstrated in Section 6.4.2. Here, however, we merely present a proof of the property for Mahalanobis's D^2.

We begin by applying to the centralized sample points $\mathbf{x}$ the whitening transformation

$$\mathbf{y} = \mathbf{k}^{-1/2}\mathbf{A}'\mathbf{x}$$

where $\mathbf{A}'$ rotates the axes to decorrelate the components of $\mathbf{A}'\mathbf{x}$ and the diagonal matrix $\mathbf{K}^{-1/2}$ scales the new axes so that the covariance matrix of $\mathbf{y}$ becomes

$$\mathbf{K}^{-1/2}\mathbf{A}'\mathbf{\Sigma}\mathbf{A}\mathbf{K}^{-1/2} = \mathbf{I}$$

For notational convenience let us put $\mathbf{K}^{-1/2}\mathbf{A}' = \mathbf{H}$. Then, in the new $(\mathbf{y})$ coordinates we have

$$\begin{aligned} D^2_{\mathbf{y}'} &= \mathbf{y}'\mathbf{y} = \mathbf{y}'\mathbf{I}\mathbf{y} \\ &= (\mathbf{H}\mathbf{x})'(\mathbf{H}\mathbf{\Sigma}\mathbf{H}')(\mathbf{H}\mathbf{x}) \\ &= \mathbf{x}'\mathbf{H}'\mathbf{H}'^{-1}\mathbf{\Sigma}^{-1}\mathbf{H}^{-1}\mathbf{H}\mathbf{x} \\ &= \mathbf{x}'\mathbf{\Sigma}^{-1}\mathbf{x} \\ &= D^2_{\mathbf{x}'} \end{aligned}$$

so that the Mahalanobis's D^2 is not changed by this transformation. (We have in fact already commented that D^2 and T^2 were chosen because they have this property.) Now in the new coordinates it is easy to see that adding another dimension can only increase $D^2_{\mathbf{y}'}$. We have

$$D^2_{(\mathbf{y}', y_d)} = (\mathbf{y}', y_d)\begin{pmatrix} \mathbf{y} \\ y_d \end{pmatrix} = \mathbf{y}'\mathbf{y} + y_d^2 \geqslant \mathbf{y}'\mathbf{y} = D^2_{\mathbf{y}'}$$

so that in general Mahalanobis's D^2 increases as the number of dimensions increases.

The practical problems of searching through the space of possible variable sets have led to the development of several approaches and it is useful to group them into three categories:

(i) Exhaustive search methods. These are only applicable when d is fairly small. The major problem is how to test the many possible sets without the large number of tests invalidating the significance level of each.

(ii) Accelerated search. We consider a branch and bound method which considers all variable sets, but does not explicitly evaluate all of them.

(iii) Suboptimal stepwise methods. These permit rapid search but at the expense of not guaranteeing to find the best solution.

6.4.1 Exhaustive search

In Section 6.4.3 we discuss stepwise methods some of which select a final variable set by sequentially adding variables and testing to see if the additions contribute significantly to the discrimination. If the order of the additions is prespecified (independently of the data) then the overall type I error rate of *all* the tests can be determined (see, for example, Roy, 1958). Unfortunately, however, the order is usually not prespecified—it is determined by the data—so that the combined type I error of all the tests is unknown. McKay (1976) presents a method to find all subsets of variables whose discriminatory power is not significantly worse than the complete variable set. The overall type I error of his method can be determined, and so can the significance level of each individual test. McKay's method combines Rao's additional information statistic (Section 6.4.3) with the union intersection principle. Readers interested in the derivation are referred to McKay (1976). First an overall test of the complete set of d variables is made, rejecting the hypothesis that the two classes are the same if

$$T_d^2 > \frac{(n_1 + n_2 - 2)dF^{\gamma}_{d,n_1+n_2-1-d}}{n_1 + n_2 - 1 - d}$$

where T_d^2 is the sample Hotelling's T^2 in d dimensions

$$T_d^2 = \frac{n_1 n_2}{n_1 + n_2}(\bar{\mathbf{x}}_1 - \bar{\mathbf{x}}_2)'\mathbf{W}^{-1}(\bar{\mathbf{x}}_1 - \bar{\mathbf{x}}_2)$$

and $F^{\gamma}_{d,n_1+n_2-1-d}$ is the γ level of the F distribution with d and $(n_1 + n_2 - 1 - d)$ degrees of freedom.

If this initial hypothesis is not rejected then the variables in the complete set do not discriminate between the classes and the analysis is finished. If, on the other hand, the hypothesis is rejected, we can go further and test subsets of variables to see if they discriminate adequately. The overall test, at the γ significance level, for this is

$$T_{d'}^2 < \frac{(n_1 + n_2 - 1 - d)T_d^2 - (n_1 + n_2 - 2)dF^{\gamma}_{d,n_1+n_2-1-d}}{n_1 + n_2 - 1 - d + dF^{\gamma}_{d,n_1+n_2-1-d}}$$

implying rejection of the hypothesis that a set of $d'(<d)$ variables is as effective as the original d. γ, it must be emphasized, is the simultaneous level—the probability that one or more hypotheses is rejected incorrectly is γ. We can go further: each hypothesis that a particular r variables are irrelevant is rejected at the α level by the above test where α is given by

$$rF^{\alpha}_{r,n_1+n_2-1-d} = dF^{\gamma}_{d,n_1+n_2-1-d}$$

A useful property of McKay's procedure is its coherence (this follows from the property of D^2 (and hence T^2) demonstrated in Section 6.4): 'All subsets of variables including adequate (γ) discriminant sets must themselves be adequate (γ) discriminant sets, and if a subset is not an adequate (γ) discriminant set then it cannot contain any adequate (γ) discriminant sets.'

The implication of this is that it is unnecessary to compute all the T_d^2 values.

McKay (1977) generalizes these two group results to the multiple class case.

To illustrate these tests we use data derived from two classes of aphids. Four measurements were used (so $d = 4$) and the class means were

$$\bar{\mathbf{x}}_1 = \begin{bmatrix} 149.09 \\ 28.91 \\ 176.34 \\ 88.74 \end{bmatrix} \quad \text{and} \quad \bar{\mathbf{x}}_2 = \begin{bmatrix} 268.10 \\ 174.71 \\ 238.10 \\ 110.94 \end{bmatrix}$$

so that

$$(\bar{\mathbf{x}}_1 - \bar{\mathbf{x}}_2) = \begin{bmatrix} -119.01 \\ -145.80 \\ -\ \ 61.76 \\ -\ \ 22.20 \end{bmatrix}$$

The within-class scatter matrix, $\mathbf{W}$ was

$$\mathbf{W} = \begin{bmatrix} 2.233 \times 10^4 & 5.748 \times 10^3 & 1.951 \times 10^3 & 3.798 \times 10^3 \\ 5.748 \times 10^3 & 9.042 \times 10^3 & 2.370 \times 10^3 & 1.025 \times 10^3 \\ 1.951 \times 10^3 & 2.370 \times 10^3 & 2.412 \times 10^4 & 1.098 \times 10^4 \\ 3.798 \times 10^3 & 1.025 \times 10^3 & 1.098 \times 10^4 & 1.616 \times 10^4 \end{bmatrix}$$

and $n_1 = 64$, $n_2 = 61$. For this data the initial overall test at the 5 per cent level is to compare

$$T_d^2 = \frac{64 \times 61}{125} (\bar{\mathbf{x}}_1 - \bar{\mathbf{x}}_2)' \mathbf{W}^{-1} (\bar{\mathbf{x}}_1 - \bar{\mathbf{x}}_2)$$

$$= \frac{64 \times 61}{125} \times 2.79 = 87.14$$

with

$$\frac{123 \times 4}{120} \times F_{4,120}^{0.05} = \frac{123 \times 4}{120} \times 2.45 = 10.05$$

The initial hypothesis that the groups are the same is rejected at the 5 per cent level.

The threshold for rejection of a subset of variables is

$$\frac{120 \times 87.14 - 123 \times 4 \times 2.45}{120 + 4 \times 2.45} = 71.27$$

The $T_{d'}^2$ values for all possible subsets are

$$
\begin{array}{lcccc}
 & \{1\;\;2\;\;3\} & \{2\;\;3\;\;4\} & \underline{\{1\;\;3\;\;4\}} & \{1\;\;2\;\;4\} \\
T_3^2 = & 76.61 & 74.24 & 24.51 & 74.55
\end{array}
$$

$$
\begin{array}{lcccccc}
 & \{1\;\;2\} & \underline{\{1\;\;3\}} & \{2\;\;3\} & \{2\;\;4\} & \underline{\{3\;\;4\}} & \underline{\{1\;\;4\}} \\
T_2^2 = & 74.55 & 23.25 & 74.14 & 73.52 & 5.04 & 19.81
\end{array}
$$

$$
\begin{array}{lcccc}
 & \underline{\{1\}} & \{2\} & \underline{\{3\}} & \underline{\{4\}} \\
T_1^2 = & 19.81 & 73.43 & 4.94 & 0.95
\end{array}
$$

The underlined sets give $T_{d'}^2 < 71.27$ so these do not discriminate as well as the complete set of four variables. It is clear from this that for this data set variable 2 plays a crucial role. From the coherence property arising from the monotonic relationship between r and T_r^2 we can see that it was unnecessary to compute $T_{d'}^2$ for sets $\{1, 3\}$, $\{3, 4\}$, and $\{1, 4\}$. These subsets are subsets of a set $\{1, 3, 4\}$ which has already been shown to be inadequate.

Gabriel (1968) has produced a similar test, except that his is a simultaneous test procedure to identify all sets of variables which discriminate significantly between subsets of groups (instead of identifying those variable sets which discriminate as well as the complete set). Again their simultaneous nature means that the overall probability of type I error is known and again they are coherent. The method uses the maximum root of $\mathbf{T}^{-1}\mathbf{B}$ as a statistic because 'the maximum root procedure is preferable to all others in that, for a given probability of any type I error, it achieves the most resolution into significant detail, that is, it provides greatest power for hypotheses or single functions of single combinations (of variables)'. For a particular variable set I and subset J of classes the statistic is

$$\theta = \text{max root of } (\mathbf{T}_{IJ}^{-1}\mathbf{B}_{IJ})$$

where $\mathbf{T}_{IJ}$ is the total sample cross-product matrix for variable set I and class set J, and $\mathbf{B}_{IJ}$ is the corresponding between-class matrix. This can be rewritten as

$$\theta = \frac{\text{max root of } (\mathbf{W}_{IJ}^{-1}\mathbf{B}_{IJ})}{1 + \text{max root of } (\mathbf{W}_{IJ}^{-1}\mathbf{B}_{IJ})} = \frac{\lambda}{1 + \lambda}$$

Critical points can be obtained from Heck's charts (see, for example, Morrison, 1967, Appendix) entered at

$$
\begin{aligned}
s &= \min(d', N - 1) \\
m &= \tfrac{1}{2}(|N - 1 - d'| - 1) \\
\tilde{n} &= \tfrac{1}{2}(n - N - d' - 1)
\end{aligned}
$$

where N is the number of classes in the subset.

If $s = 1$ (for example, if $N = 2$ and $d' \geqslant 2$) then the transformation

$$L = \frac{\tilde{n} + 1}{m + 1} \cdot \lambda$$

gives a variate L which is distributed as an F variate with $(2m + 2)$ and $(2\tilde{n} + 2)$ degrees of freedom.

Using the data of the preceding example let us investigate whether the set of the first two variables, $I = \{1, 2\}$, discriminates adequately for the subset of classes consisting of $J = \{\text{class 1, class 2}\}$ (i.e. all the classes in our data).

We have

$$\mathbf{W} = \begin{bmatrix} 2.233 \times 10^4 & 5.748 \times 10^3 \\ 5.748 \times 10^3 & 9.042 \times 10^3 \end{bmatrix}$$

$$\mathbf{W}^{-1} = \begin{bmatrix} 5.353 \times 10^{-5} & -3.403 \times 10^{-5} \\ -3.403 \times 10^{-5} & 1.322 \times 10^{-4} \end{bmatrix}$$

$$\mathbf{B} = 31.232 \times \begin{bmatrix} 1.416 \times 10^4 & 1.735 \times 10^4 \\ 1.735 \times 10^4 & 2.126 \times 10^4 \end{bmatrix}$$

$$\mathbf{W}^{-1}\mathbf{B} = 31.232 \times \begin{bmatrix} 0.1676 & 0.2053 \\ 1.812 & 2.22 \end{bmatrix}$$

Solving $|\mathbf{W}^{-1}\mathbf{B} - \lambda\mathbf{I}| = 0$ yields $\lambda = 75$. Now

$$s = \min(2, 1) = 1$$
$$m = \tfrac{1}{2}(|2 - 1 - 2| - 1) = 0$$
$$\tilde{n} = \tfrac{1}{2}(125 - 2 - 2 - 1) = 120$$

Thus

$$L = (121/1) \times 75 = 9{,}075$$

Since the 5 per cent level for the F distribution with $(2m + 2, 2\tilde{n} + 2) = (2, 242)$ degrees of freedom is 3 this is a very highly significant result.

Other authors have studied methods of calculating the criterion values for each member of the possible large set of variable subsets. McCabe (1975), for example, adopts the algorithm of Furnival (1971) (developed for regression) to the problem of computing $|\mathbf{W}|/|\mathbf{T}|$ for all subsets of d' variables.

6.4.2 Accelerated search

In Chapter 2 we considered a branch and bound method for accelerating the search for nearest neighbours and in Chapter 7 we apply such an approach to search for an optimal partition of sample points. Here we use the method to find the best d'-dimensional subset from the complete set of d dimensions. Kittler (1978a) describes an algorithm based on this method. Since the method can be implemented in different ways we shall be content to give an abstract outline of the approach. The only condition necessary for the method is the monotonicity condition described in Section 6.4.

Beginning from the set consisting of all d available variables one can construct a tree structure by successively deleting variables. Thus, for example, from the node corresponding to the complete variable set, d new nodes can be generated, each one having a single variable deleted. Then, from each of these nodes, $(d - 1)$ further nodes can be generated. And so on until the level is reached at which all nodes have d' variables.

This description represents a particular way of constructing the tree but others could be used. The only constraints are that all final nodes should contain d' variables and that as one proceeds down the tree (from the single original node to one of the many final nodes) variables are always discarded and never added. It is also necessary, of course, that the set of final nodes should consist of all possible sets of d' from d variables. (Fortunately the tree does not need to be generated explicitly—one needs only to consider a single branch at a time.) We shall suggest below a tree structure which leads to a more efficient search.

Now suppose that some branch of the tree has been followed to its final node and a criterion value, J_1, has been computed for that node. That is to say, the separability criterion has value J_1 for the variable set represented by this node. If we then proceed down some other branch of the tree and find at some node that the criterion J_2 is less than J_1 there is no point in continuing further down this branch. The monotonic property of J means that the criterion can only decrease yet further—the final nodes of this branch will also have J-values less than J_1—and our aim is to find that final node with the largest J-value. We can thus stop the downward search and back up the current branch to the previous node. From here we take a branch which has not yet been searched and continue down it until either we reach a terminal node or again J becomes less than J_1. Whenever we back up to a node of which all branches have been searched we continue further up the tree until we discover one with unsearched branches. If we ever reach a terminal node which has a J-value larger than J_1 this value replaces J_1 for future stages of the search.

Although the tree structure described above would work it is not very efficient. One reason for this is that the very simple tree above repeats all the variable sets (for example it includes sets generated by discarding v_i followed by v_j as well as all sets generated by discarding v_j followed by v_i). These unnecessary repetitions can be deleted in various ways and one way which will increase the search efficiency is based on the fact that a section of the tree branching from a node with a low J-value is more likely to be discarded than if the node has a high J-value. The freedom of choice in the tree structure can thus be used to make more branches emerge from nodes with low J-values.

To illustrate these theoretical ideas again consider the data of the examples above. Suppose we wish to find the best two-variable subset—we can establish the tree structure of Figure 6.8. The numbers in brackets indicate the set of variables being considered and the other members are T^2 for these sets.

We begin by evaluating nodes {3, 4}, {2, 4}, and {2, 3} and choosing {2, 3} since it has the largest J-value (74.14). From there we back up the tree until we come to a node for which not all the branches have been evaluated.

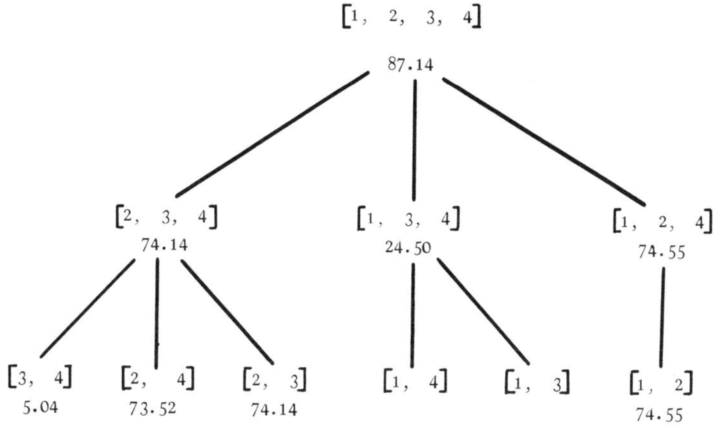

Figure 6.8 A branch and bound search to find the best two-variable subset of the four variables of the aphid data. [$x, y, \ldots$] signifies the set of variables $x, y, \ldots$, etc. The numbers underneath are T^2 ($=J$)-values

{1, 2, 3, 4} is the first such node. Proceeding down to node {1, 3, 4} we find $J = 24.5$. Now, since J can only decrease as we proceed down the tree and we already have node {2, 3} with a value larger than 24.5 there is no point in evaluating nodes {1, 4} and {1, 3}. We thus back up the tree to {1, 2, 3, 4} and proceed down the final unevaluated branch to {1, 2, 4}. Here $J = 74.55$ which is greater than 74.14 so we cannot reject the final nodes associated with {1, 2, 4}. Evaluating {1, 2} we find $J = 74.55$, the largest value for any two-variable subset.

The branch and bound algorithm has found many applications throughout statistics (see, for example, Hand, 1981) and operational research and often makes practicable problems for which exhaustive search would be totally out of the question. Usually these are problems where the number of possibilities which must be examined increases exponentially with some fundamental parameter of the problem. Unfortunately, although the branch and bound algorithm slows down the exponential growth rate of the number of possibilities which must be evaluated, it still remains exponential. It is thus sometimes necessary to resort to suboptimal search methods.

6.4.3 Suboptimal search methods

One obvious way to find a subset of d' from d variables would be to take the d' individually best. We have already commented, however, that this need not result in a good set of variables since it completely ignores multivariate relationships. Take the simple case of normally distributed variables. The set of d' individually best variables might be little better than a single variable if they are highly correlated. One would do better to choose d' uncorrelated variables, on the grounds that each adds something the others have not got, even though each as an individual does not discriminate very well between the classes.

An example illustrating the necessity of taking multivariate relationships into account occurred while developing the questionnaire mentioned in Section 5.2.2. The two individually best questions (chosen from 140) had misclassification rates of 8.00 and 8.04 per cent and when taken together a combined misclassification rate of 5.2 per cent. However, the best *pair* of questions (chosen by comparing all pairs—this is possible since there are only $\binom{140}{2} = 9730$ of them—but one can see how rapidly the number of subsets of size d' increases as d increases), had individual misclassification rates of 10.16 and 8.04 per cent and a combined rate of 4.2 per cent. (The reader should note that these error rates refer to the design set, which explains why they are much smaller than in Section 5.2.2.)

Liddell (1977) gives another example: four measurements on the volume–pressure curves of lung functions are used to compare a group of female smokers with those of female non-smokers. The t-statistics between the two groups for each measurement considered alone are given in Table 6.1. Measures 1 and 2 produced the largest (in absolute terms) t-statistics. However, when taken in pairs, measures 1 and 2 yielded the smallest T^2.

From the above examples one might suppose that dependence and small sample-size-to-dimensionality ratios were the sole causes of the phenomenon. In fact there are cases when the d' best variables are not the best d' variables even when the variables are independent and an infinite set of observations is available. Take, for example, the illustration of Elashoff *et al.* (1967). This has three binary variables v_1, v_2, and v_3, with

$$P(v_1 = 1/\omega_2) = 0.02 \qquad P(v_1 = 1/\omega_1) = 0.95$$
$$P(v_2 = 1/\omega_2) = 0.04 \qquad P(v_2 = 1/\omega_1) = 0.90$$
$$P(v_3 = 1/\omega_2) = 0.01 \qquad P(v_3 = 1/\omega_1) = 0.80$$

If the prior probabilities are equal it is easy to show that the misclassification

Table 6.1 t-Statistics between female smokers and female non-smokers on each variable separately, and T^2-statistics using pairs of variables

	Measure 1	Measure 2	Measure 3	Measure 4
t-statistics	−2.59	−1.83	−1.60	−0.67
		T^2-statistics		
Measure 1	—	0.67	6.82	6.96
Measure 2		—	3.40	5.21
Measure 3			—	2.60

(This table is reproduced from *The Statistician*, vol. 26, No. 1, p. 5 with permission from the editor.)

rate for v_1 alone (0.07) is less than that for v_2 alone (0.14) which in turn is less than that for v_3 alone (0.21). However, when it comes to choosing pairs of variables the misclassification rate of (v_1, v_3) (0.04) is less than that of (v_1, v_2) (0.06). This is despite their independence. Cover (1974) gives another example where two independent measurements of the worst variable are better than one measurement on the best combined with one on the worst, which in turn are better than two independent measurements on the best variable.

These examples illustrate problems which can be encountered if no attempt is made to allow for the multivariate nature of the data. In between the two extremes of exhaustive search and individual selection lie a number of sequential methods. The following two algorithms lie at the base of such methods:

(i) *Sequential forward selection.* Suppose that d_1 variables, forming a set x_{d_1}, have already been selected. Now, for each variable $v_r \in \{x_d - x_{d_1}\}$ examine the variable set $\{x_{d_1} \cup v_r\}$. (x_d here is the complete variable set.) This produces $(d - d_1)$ criterion values, $J(x_{d_1} \cup v_r)$. The v_r which produces the maximum criterion is then chosen and added to x_{d_1} to give set x_{d_2}. The whole process is then repeated, examining each variable in the set $\{x_d - x_{d_2}\}$ to see which, when added to x_{d_2}, gives the greatest J-value. At each stage we add the single variable which, in combination with those already selected maximizes the criterion.

This approach has two disadvantages: first no account is taken of the interrelationships between elements of the set $\{x_d - x_{d_1}\}$; that is, between variables which have not yet been selected. Secondly, once a variable has been chosen and added to x_{d_1} there is no way it can be removed should subsequent additions render it unnecessary.

(ii) *Sequential backward elimination.* The principle of this approach is much the same as that of the preceding one except that variables are deleted one at a time instead of being added. The process thus begins with the complete set and removes that variable v_r which decreases J the least. That is, the v_r is selected for which $J(x_d - v_r)$ is greatest. Then each remaining variable v_s is examined and that one removed which gives the greatest $J(x_d - v_r - v_s)$. And so on.

This method has the disadvantage that it is computationally more demanding than forward selection. It does, however, have the advantage that $J(x_d)$ is evaluated so that the system's performance (in terms of J) can be judged relative to the performance on the complete variable set.

Relationships between the discarded variables are ignored and once a variable has been discarded it cannot be re-added.

Using the sequential forward selection method on the data used above we first find that single variable which has the maximum criterion. In this case (using T^2 as criterion) v_2, with $T^2 = 73.43$, is chosen. Now we evaluate T^2 for the variable sets $\{v_2, v_1\}$, $\{v_2, v_3\}$, and $\{v_2, v_4\}$. Set $\{v_2, v_1\}$ has the largest $T^2 = 74.55$. Then we consider sets $\{v_2, v_1, v_3\}$ and $\{v_2, v_1, v_4\}$, selecting the first because it has the larger $T^2 = 76.61$.

Using sequential backward elimination we begin with set $\{v_1, v_2, v_3, v_4\}$ yielding $T^2 = 87.14$, and investigate the four sets resulting from deletion of a single variable

$$\begin{array}{ll} \{v_1, v_2, v_3\} & T^2 = 76.61 \\ \{v_1, v_2, v_4\} & T^2 = 74.55 \\ \{v_1, v_3, v_4\} & T^2 = 24.51 \\ \{v_2, v_3, v_4\} & T^2 = 74.24 \end{array}$$

Clearly deletion of v_4 results in the smallest reduction of the criterion. We thus go on to consider sets $\{v_1, v_2\}$, $\{v_2, v_3\}$, and $\{v_1, v_3\}$.

Although in this example the forward and backward methods lead to the same results, this need not always be the case.

Kittler (1978a) outlines various extensions of these techniques, the most general of which is his 'generalized plus l—take away r selection algorithm'. Each step finds the particular l-dimensional subset of those variables not yet added which, when combined with the currently selected set, leads to the greatest J. Then each step examines the selected set to identify those r variables which, when discarded, reduce J by the least. Thus, the process iterates through steps of adding l and taking r variables. If l is greater than r the process is bottom up, and if l is less than r it is top down. If l or r is greater than 1 some of the disadvantages mentioned above are alleviated. Computational requirements prevent l and r from being too large, however.

Backer and De Shipper (1977) present a sequential algorithm based on the following. Let $x_{d_1} = \{v_r | r = 1, \ldots, d_1\}$ the set of variables which have already been chosen and let $y_{d_1} = \{x_d - x_{d_1}\}$, i.e. y_{d_1} is the set of variables not yet chosen. Backer and De Shipper's max–min algorithm assumes that we have available criterion values for each variable individually ($J(u)$; $u \in x_d$) and for each pair of variables ($J(u, v)$; $u, v \in x_d$). Thus, we can see if $u \in y_{d_1}$ contributes anything to the set already chosen by considering the values of

$$\Delta J(v, u) = J(v, u) - J(v)$$

for each $v \in x_{d_1}$. The proposed new element u will be of little value if ΔJ is near zero. In such a case u will add little information to that provided by v. Backer and De Shipper thus select as the next element that u which maximizes

$$\min_{v \in x_{d_1}} \Delta J(v, u)$$

In full, choose the $u \in y_{d_1}$ which gives

$$\max_{u \in y_{d_1}} \left(\min_{v \in x_{d_1}} \Delta J(v, u) \right)$$

The method is a compromise between computational feasibility and multivariate relationships—all relationships involving more than two variables are ignored. Note also that the method is at risk from the sort of problem illustrated by Cover (1974) and referred to above.

Kittler (1978a) reports some empirical comparisons between various methods. His data consisted of two normal classes in a 20-dimensional space and he used Mahalanobis's D^2 as the separability measure. His general conclusions were that forward selection and backward elimination methods selecting (or rejecting) several variables simultaneously were better than methods which select or reject only one variable at each step. This is hardly surprising, and neither is the fact that methods which select several variables simultaneously require more computer time. He also found that a combination of forward selection of two variables and backward deletion of one variable almost always gave optimal results and compared in time very favourably with the branch and bound method of Section 6.4.1. More general plus l—take away r algorithms, however, only gave good results when $l \simeq r$. The max–min algorithm was found to be comparatively poor, but again this is not surprising since it uses far less information. The decision about which method to use would normally be based on a compromise between expense (in terms of computer time) and flexibility (in terms of how well the method caters for multivariate relationships).

We have now discussed ways for determining which variable is the best to select next and thus far there is nothing to prevent us from continuing until we have selected as many as our computational facilities can comfortably handle. However, since we know the risks associated with this it would be useful if we had some means of deciding whether an extra $(d - d')$ variables make a significant contribution to the discrimination. Rao (1970) gives a statistic to test this. He uses

$$F \sim \frac{(n - d - 1)n_1 n_2 (D_d^2 - D_{d'}^2)}{(d - d')[n(n - 2) + n_1 n_2 D_{d'}^2]}$$

where n is the total sample size and where D_d^2 is Mahalanobis's D^2 evaluated on d variables and $D_{d'}^2$ is the same evaluated on the subset of d' variables. This statistic has an F distribution with $(d - d')$ and $(n - d - 1)$ degrees of freedom. However, as McKay (1976) points out and as we have already noted in Section 6.4.1, unless the order of selecting the variables is determined independently of the data, the overall error rate of a sequence of such tests is unknown.

Costanza and Afifi (1979) have used Monte Carlo studies to compare the best subsets chosen by different stopping rules. Their comparison is on the basis of conditional and estimated unconditional probabilities of correct classification.

6.5 SELECTION BY TRANSFORMATION

So far we have considered selection of a subset from the complete set of variables. This sort of approach would be appropriate when there is some reason for wishing to avoid making as many measurements on future objects as were made on the design set—cost could be such a reason. When it is quite feasible to measure the original large set, but for other reasons (such as accuracy) it is required to reduce the dimensionality of the classifier, transformation methods can be adopted. Such methods find a set of axes spanning that subspace of the complete space in which class separability is maximized. In common with much of the rest of multivariate statistical analysis most methods used linear transformations, and most of these are based on scatter matrices. As a result of this the mechanics of the methods are implicit in our earlier discussions of separability criteria.

First let us suppose that there are only two classes and that these are completely determined by the class means $\bar{\mathbf{x}}_i (i = 1, 2)$ and the common within-class matrix $\mathbf{W}$. This is the situation tackled by classical discriminant analysis, which seeks to find that single dimension $\mathbf{v}'\mathbf{x}$ along which the criterion

$$J = \frac{\text{between-class spread}}{\text{within-class spread}} = \frac{\mathbf{v}'\mathbf{B}\mathbf{v}}{\mathbf{v}'\mathbf{W}\mathbf{v}} \tag{1}$$

is maximized. Differentiating J with respect to $\mathbf{v}$ and equating to zero gives

$$(\mathbf{v}'\mathbf{W}\mathbf{v})(2\mathbf{B}\mathbf{v}) = (\mathbf{v}'\mathbf{B}\mathbf{v})(2\mathbf{W}\mathbf{v})$$

and thus

$$(\mathbf{W}^{-1}\mathbf{B} - J\mathbf{I})\mathbf{v} = 0$$

So the optimal $\mathbf{v}$ is an eigenvector of $\mathbf{W}^{-1}\mathbf{B}$ with J its corresponding eigenvalue. For two classes $\mathbf{B}$ is of rank 1 so $\mathbf{W}^{-1}\mathbf{B}$ has only one eigenvalue. We can therefore write

$$\begin{aligned} J = \operatorname{tr} \mathbf{W}^{-1}\mathbf{B} &= \frac{n_1 n_2}{n_1 + n_2} \operatorname{tr} \mathbf{W}^{-1}(\bar{\mathbf{x}}_1 - \bar{\mathbf{x}}_2)(\bar{\mathbf{x}}_1 - \bar{\mathbf{x}}_2)' \\ &= \frac{n_1 n_2}{n_1 + n_2} (\bar{\mathbf{x}}_1 - \bar{\mathbf{x}}_2)'\mathbf{W}^{-1}(\bar{\mathbf{x}}_1 - \bar{\mathbf{x}}_2) \\ &= T^2 \end{aligned}$$

If there are $N (N > 2)$ classes then B has rank $r = \min(d, N - 1)$ and r eigenvectors (termed 'canonical variates') can be extracted from $\mathbf{W}^{-1}\mathbf{B}$. These eigenvectors span that r-dimensional subspace in which the N classes are most separated, as measured by criterion J.

In the two-class case the criterion $|\mathbf{W}|/|\mathbf{T}|$ leads to the same result.

This generalization of Fisher's approach is known as canonical variate analysis and is the discriminant analysis method most widely implemented in computer packages. Often it is applied to the selection of a subset of the original variables rather than to dimensionality reduction via transformation. This is done by using the standardized discriminant function coefficients (i.e. the coefficients calculated from standardized data) as indicators of the importance of the original variables. Thus, a large coefficient (positive or negative) means that the corresponding variable makes an important contribution to the between-group separability.

By studying (1) above it is easy to see that high within-group correlation of opposite sign to the between-groups correlation will lead to a greater criterion value than when the within-groups correlation is low or of the same sign as the between-groups correlation. However, Campbell (1980) points out, by analogy with regression when the predictors are correlated, that such a situation may lead to unstable estimates of the standardized discriminant function coefficients. He then outlines a method, analogous to ridge regression, for alleviating the problem. This also suggests another criterion for rejecting variables—if they have unstable discriminant function coefficients they may be redundant.

One point which should be emphasized is that throughout the above we have assumed that the means and within-class scatter matrices completely determined the distributions, and that these within-class scatter matrices were identical. If they are not identical then the optimal decision surface is not linear—so a linear transformation will not be optimal (but a quadratic one will). This extends to the more general case when the assumption that the $\boldsymbol{\mu}_i$ and $\boldsymbol{\Sigma}_i$ adequately represent the distributions is unjustified. In such a case the optimal transformations are non-linear.

The preceding point emphasizes the somewhat artificial nature of the separation into variable selection and classifier design. If a transformation is found such that in the new space the optimal decision surface is linear then classifier design is trivial. The (possible highly complex) variable selection transformation has done the work of the decision surface. However, one could split the work more equitably, using a simpler transformation and a more flexible decision surface. This is the approach usually adopted, using a linear transformation to identify an effective subspace and then a flexible decision surface in this subspace.

In this section we have discussed ways of finding the best discriminating subspace spanned by $\min(d, N-1)$ dimensions. It is clear that if $\min(d, N-1) = d$ then no dimensionality reduction has been achieved. Even if this is not the case $(N-1)$ dimensions might be too many. If this is true we can, of course, simply choose the first d' eigenvectors and hope that they provide sufficient discrimination. A statistic which tests this hope is

$$\chi^2 = -\left(n - \frac{d+N}{2} - 1\right) \ln \Lambda'$$

(where

$$\Lambda' = \prod_{j=d'+1}^{d} \frac{1}{1+\lambda_j}$$

with λ_j being the jth eigenvalue of $\mathbf{W}^{-1}\mathbf{B}$), which is distributed as χ^2 with $(d - d')(d - d' - 1)$ degrees of freedom. This can be used as a test of significance of the discrimination afforded by the last $(d - d' + 1)$ eigenvectors after the first d' have been used.

An interesting paper in this area is that by Fukunaga and Ando (1977) who consider non-linear feature extraction using the discriminant analysis separability criterion tr $\mathbf{T}^{-1}\mathbf{B}$. They show that the posterior probabilities $P(\omega_i|\mathbf{x})$ are the optimum non-linear features with respect to this criterion (as well as being the optimum features in the Bayes sense). They also show that the difference between the optimum value of the criterion tr $\mathbf{T}^{-1}\mathbf{B}$ and the criterion value in a space spanned by an approximating set of variables is the mean square error between the selected features and the optimum features $P(\omega_i|\mathbf{x})$.

6.6 FURTHER READING

At the time of writing there appears to be no single textbook devoted to the problem of variable selection. There are, however, chapters in various textbooks—for example, Meisel (1972) has such a chapter with an associated bibliography of 104 references on variable selection in pattern recognition. Other surveys are those of Kittler (1975), one of the most comprehensive ones available, and Kittler (1978a), the latter presenting a comparison of methods for selecting subsets of variables.

Ali and Silvey (1966) present a summary of some of the work of separability measures, and Goldstein and Rabinowitz (1975) consider variable selection for two classes having multinomial distributions. Goldstein and Dillon (1978) contains a chapter on variable selection for discrete variables. Decell and Guseman (1979) provide a discussion of recent results on separability criteria and optimization algorithms. McKay (1978) presents a graphical variable selection method derived from a method for variable selection in regression. This is a convenient point to note that a considerable amount of effort has been expended on variable selection in regression (see, for example, Beale, 1970, and Draper and Smith, 1966)—and that because discriminant analysis can be considered as a special case of regression there is similarity between the methods discussed above and those used in regression.

Wahl and Kronmal (1977) compare the performance of the usual linear decision surface method of Fisher (see Section 4.5) with that of the quadratic decision surface method which results when the variance–covariance matrices are not assumed equal. Since such a quadratic surface can be interpreted as linear in a space spanned by up to second-order terms in the original variables (as explained in Section 4.1) this is a comparison of two different variable sets, one a subset of the other. The results of Wahl and Kronmal show that the design set size is a crititical factor in choosing between the two approaches.

One situation which has not been mentioned is that occurring when there is an extremely large original set of variables. Such problems can sometimes occur—even to the extent that $d > n$. Jain and Dubes (1978) have studied such problems and formulate a two-stage process:

(i) Divide the variables into mutually exclusive subsets such that variables in the same subset are similar to each other and different from those in other subsets.

(ii) Reduce each subset to a single representative variable.

Conventional variable selection algorithms can then be applied if further reduction is desired.

The first stage can be by means of a cluster analysis approach and the second can either be a selection process (choosing a single existing variable from each subset) or a transformation process (e.g. the largest principal component).

EXERCISES

6.1 Generate 50 points (x_i, y_i), where x_i lies in the range 0 to 1 and

$$y_i = a + bx_i + cx_i^2 + \varepsilon$$

with a, b, and c fixed coefficients and ε a standardized normal random variate. The true regression line is thus

$$y_i^* = a + bx_i + cx_i^2.$$

Use a standard regression package to fit linear, quadratic, and cubic regression line, $\hat{y}_i = \hat{y}_i(x)$, to the points. Now calculate the sum of squares fits

$$\sum_{i=1}^{50} (y_i^* - \hat{y}_i)^2$$

and

$$\sum_{i=1}^{50} (y_i - \hat{y}_i)^2$$

for each of the three regression lines. How do the results relate to the discussion of dimensionality and misclassification rate in Section 6.2?

6.2 An approach to separability measures not discussed in the text is based on the proportion of design set elements whose nearest neighbour is from the same class. Discuss operational measures derived from this approach.

6.3 Write a computer program to investigate all possible subsets of variables using McKay's method (Section 6.4.1).

6.4 For an arbitrary set of elements $\{a, b, \ldots, c\}$ devise a criterion function J which yields a value for each subset and which obeys the monotonicity condition of Section 6.4. Write a computer program to find the optimum subset, as determined by J, using the branch and bound algorithm (Section 6.4.2). Compare the number of subsets on which J was evaluated with the number which would need to be evaluated in an exhaustive search.

6.5 Devise a data set for which the sequential forward selection method of Section 6.4.3 would miss the optimum subset of variables.

6.6 Most existing computer discriminant analysis packages are based on the classical canonical variates method (Section 6.5). Output from such programs often contains standardized discriminant function coefficients, i.e. the discriminant functions derived from the data after it has been standardized (to zero mean and unit variance). One way of selecting variables on the basis of these standardized functions would be to choose those variables with the larger coefficients. What are the advantages and risks associated with such a procedure?

6.7 Show that for two classes the separability criteria tr $\mathbf{W}^{-1}\mathbf{B}$ and $|\mathbf{W}|/|\mathbf{T}|$ are equivalent.

CHAPTER 7

Cluster Analysis

7.1 INTRODUCTION

In Section 1.1 two types of problems were defined. So far only one of these, that of devising classification rules from a set of classified objects, has been considered. We now turn to the second type, that of cluster analysis, where a sample of objects of unknown classification is available and the aim is to group those objects into natural classes or clusters.

The fact that there is no *a priori* classification of the sample suggests that cluster analysis is fundamentally a tool for data exploration. That is to say, one wishes to study the data to see if natural and useful groupings do in fact exist. Although this is by far the most important circumstance under which cluster analysis techniques are used, there are other circumstances. For example, the cost of acquiring an initially classified sample might be too great, or perhaps the class structure is known to vary with time. In any case, whatever the motive behind the application, the first thing to note is that there are many possible classifications which can be imposed on a sample. Thus, for example, human beings can be classified as male or female, by social class, by skin colour, by educational attainment, by age. The list is endless and it is apparent that the sort of grouping which emerges from an analysis will depend very much on the variables used to represent the object. This is a crucial point since poor choice of variables can lead to a clustering which is useless for a particular purpose. Consider, for example, an attempt to see if people fall naturally into well-defined reading ability groups. Then there would be little point in using height and hip-to-waist ratio as variables since a natural grouping into sexes is much more likely to result.

Having decided which variables to measure it may be necessary to reduce the number to make computation feasible or to eliminate ineffective variables. In the preceding chapter we considered this problem for the case of distinct groups of classified observations. There the aim was to reduce the dimensionality while maintaining class separability. This meant that we could use a measure of class separability as a criterion by which to gauge the value of the dimensionality reduction. In the present case, however, we cannot use such a measure since the groupings are *a priori* unknown. Consequently, the aim must be the more general one of reducing dimensionality while maintaining the *structure* of the observations as far as is possible. That is, we seek a space of lower dimension than the original one which approximates the original space distribution of observation points as accurately as possible. To achieve this we need two things:

(a) a measure showing how closely the subspace representation matches the original representation, and

(b) an algorithm for finding that subspace which optimizes the measure.

For (b) early work concentrated on linear transformations followed by deletion of the least important new variables. In this group we have what is doubtless the most widely used approach: the method of principle components. This identifies and discards those (mutually orthogonal) directions in the original space which account for very little of the variance of the sample. By discarding these, little is sacrificed in terms of accuracy of description of the sample's structure.

When the subspace identification problem is expressed in the form of finding dimensions which minimize (or, by complementarity, maximize) spreads it becomes apparent that other criteria could be used. We could choose the d'-dimensional hyperplane which maximizes the sum of squared between sample distances

$$\sum_i \sum_j (\mathbf{x}_i - \mathbf{x}_j)'(\mathbf{x}_i - \mathbf{x}_j)$$

or the entropy function measuring the uniformity of projections onto the axes of the hyperplane

$$-\sum_{i=1}^{d'} p_i \log p_i$$

(where p_i is the sample mean of the squared projections of the $\mathbf{x}_j$ onto the ith new co-ordinate). See, for example, Tou and Heydorn (1967) and Watanabe *et al.* (1967). In fact many of these criteria lead back to selecting the eigenvectors associated with the d' largest eigenvalues of the sample covariance matrix, i.e. the principal components solution. One other point to note is that in the pattern recognition literature the principal components method is usually referred to as the Karhunen–Loève transformation.

It has been suggested (see, for example, Bartko *et al.*, 1971) that unless there is a pronounced cluster structure the results of a cluster analysis applied to the original variables can differ markedly from a similar analysis applied to the space spanned by the first few components. This is, of course, always a risk: however one looks at it, dimensionality reduction does mean a sacrifice of data.

More recently, non-linear methods have been developed. Non-linear transformations of the sample from the d-dimensional space to the lower d'-

dimensional space have been based on several structural criteria. Amongst the most popular are Kruskal's 'stress' (Kruskal, 1964) defined by

$$\sum_{i<j} (D_{ij} - D'_{ij})^2 \Big/ \sum_{i<j} D_{ij}^2$$

(where D_{ij} is the distance (or dissimilarity–see below) between objects $\mathbf{x}_i$ and $\mathbf{x}_j$ in the d-dimensional space and D'_{ij} is the corresponding distance in the d'-dimensional space) and Sammon's criterion

$$\frac{1}{\sum_{i<j} D_{ij}} \sum_{i<j} (D_{ij} - D'_{ij})^2 / D_{ij}$$

A recent survey of such 'multidimensional scaling' techniques is presented by Shepard *et al.* (1972), and a descriptive survey of problems and how they might be overcome by Shepard (1974).

Whether a linear or non-linear method is adopted one can either project to a $d'(<d)$-dimensional space which preserves structure fairly well and then perform a cluster analysis in this new space or one can go to the extreme of $d' = 2$. In such a case inspection by eye is possible—and may be very revealing.

Because of its importance as the most popular linear data reduction technique for unclassified data, it should be pointed out that the principal components method has disadvantages. It is, for example, not invariant to changes of scale of the original variables. This is in contrast to the classified case of Chapter 6 where class separability is invariant under any non-singular transformation.

So far no attempt has been made to define the subject of this discussion, the word 'cluster'. The reason for this is simply that what is an appropriate definition for one application may not be so for another. In some cases connected regions of relatively high probability density may constitute clusters, whereas in others only compact roughly hyperspherical groups of objects may be eligible. It is generally true that ultimately the investigator must decide what he means by cluster, and care needs to be taken that an artificial structure is not imposed on the data. This fact of multitudinous definitions is one of the reasons behind the large number of cluster analysis techniques which have been developed (another is that techniques have been developed by workers from a very wide variety of fields). Different methods can lead to different shapes of cluster, though clearly if the data has a strong structure one would hope that this would be detected by most techniques. Indeed, it has been recommended that one way to test structure validity is to apply several techniques to the data.

Sections 7.3 and 7.4 deal with cluster analysis techniques themselves. The division into two types follows the common and convenient pattern, distinguishing between hierarchical methods and optimization methods. However, in the expositions it is usually assumed that c, the number of clusters, is specified *a priori*. Sometimes such an assumption is quite reasonable: for example, if we began with a small classified sample (perhaps the cost of classification is very great) and wish to make our statistics more representative of the parent population by using a large unclassified sample. Sometimes a particular division into a number of clusters is not wanted at all, but rather a hierarchy of subclusters within clusters is desired (as, for example, in tracing evolutionary trees). The hierarchical techniques of Section 7.3 are appropriate here.

Often, however, we are simply given a sample of data and have no justification for choosing a particular value for c. Finding c becomes part of the clustering problem. A general approach is to compare some criterion evaluated for different values of c. One possible, and at first intuitively appealing, criterion is the average within-cluster distance (Thorndike, 1953). The hope is that a plot of this against c will fall with increasing c until the optimum value is reached, at which point the curve will suddenly flatten out. Unfortunately, it seems that this intuitive ideal is not achieved unless the clusters are compact and well separated.

Following similar arguments, the criterion $|\boldsymbol{\Sigma}|$ (where $\boldsymbol{\Sigma}$ is the within-cluster covariance matrix) has been suggested. Marriott (1971) has modified this to $c^2|\boldsymbol{\Sigma}|$, suggesting that the c-value at which this is a minimum will give the optimum number of clusters. He also points out that $c^2|\boldsymbol{\Sigma}|$ should remain fairly constant for a uniform distribution.

If it seems justifiable to make assumptions about the forms of the cluster distributions, then it might be possible to proceed in the classical statistical way of comparing an obtained criterion value with the distributions expected under a null hypothesis specifying a certain number of clusters. Everitt (1981), working along these lines, has produced a test of a single multivariate normal distributions against multiple clusters.

7.2 DISTANCE MEASURES

The difficulties associated with the question of whether or not to standardize variables were pointed out in Chapter 1. In cluster analysis these difficulties can be particularly severe since different sets of weights on the variables can lead to completely different apparent structures. An illustration is given in Figure 7.1. Figure 7.1(a) shows the unweighted data and 7.1(b) shows weights applied such that $X = x/10$ and $Y = 10y$.

Standardization is just one of the problems associated with cluster analysis. Many others are specific to particular techniques and will become apparent below, but one which is general is the difficulty of deciding which measure of distance between points to use. We observed in Chapter 1 that measures other than the usual Euclidean distance

$$d_1(\mathbf{x}, \mathbf{y}) = \left[\sum_{i=1}^{d} (x_i - y_i)^2\right]^{1/2}$$

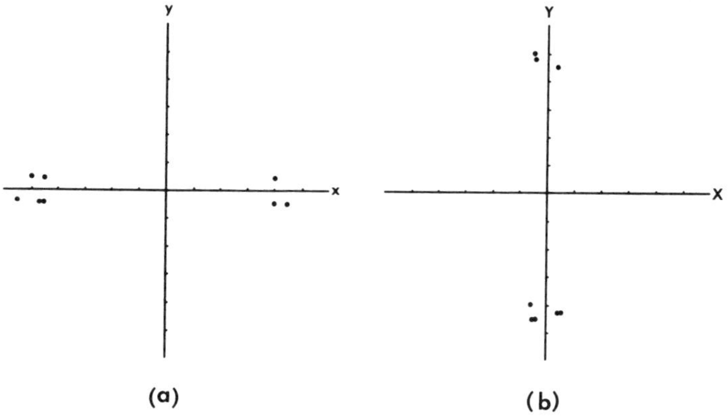

Figure 7.1 Reweighting variables can lead to quite different cluster structures

existed and that the use of d_1 was not mandatory. Often, in fact, other measures are more suitable. In this section we outline some alternatives, beginning with measures for interval scales.

Perhaps the first thing to note is that in many problems different variables are not strictly comparable. If blood-sugar level and height constitute two of the variables does it make much sense to combine them into a single measure of distance? In the language of the primary school, it is rather like adding apples to oranges. In trying to formulate effective distance measures we are trying to combine the different variable types into a single meaningful scale. One measure which perhaps makes more intuitive sense in its way of combining different variable types is the *city block* distance

$$d_2(\mathbf{x}, \mathbf{y}) = \sum_{i=1}^{d} |x_i - y_i|$$

Now objects which are 3 units apart on one variable and 1 unit apart on another have the same distance between them as objects which are 2 units aparts on each variable.

Both d_1 and d_2 are special cases of the general *Minkowski* distance

$$d_3(\mathbf{x}, \mathbf{y}) = \left[\sum_{i=1}^{d} |x_i - y_i|^r \right]^{1/r}$$

When $r = 2$ we have d_1, and when $r = 1$ we have d_2. Another special case which is sometimes used occurs when $r = \infty$

$$d_4(\mathbf{x}, \mathbf{y}) = \max |x_i - y_i|$$

The Minkowski distance is one way in which Euclidean distance could be generalized. Another is

$$d_5(\mathbf{x}, \mathbf{y}) = (\mathbf{x} - \mathbf{y})'\mathbf{A}(\mathbf{x} - \mathbf{y})$$

where $\mathbf{A}$ is a $d \times d$ matrix. d_1 occurs when $\mathbf{A} = \mathbf{I}$. If $\mathbf{x}$ and $\mathbf{y}$ are the means of two distributions and $\mathbf{A}$ is the inverse of their (common) variance–covariance matrix, then d_5 is a measure of the distance between the two distributions. It is known as the Mahalanobis distance, denoted Δ^2, and we have already met it in Chapter 6. With $\mathbf{A}$ diagonal, d_5 is equivalent to using Euclidean distance on rescaled axes.

Readers may sometimes come across the term *metric* used to describe a distance measure. A metric is a measure which satisfies

(1) $d(\mathbf{x}, \mathbf{y}) \geqslant 0$; $d(\mathbf{x}, \mathbf{x}) = 0$
(2) $d(\mathbf{x}, \mathbf{y}) = d(\mathbf{y}, \mathbf{x})$
(3) $d(\mathbf{x}, \mathbf{y}) \leqslant d(\mathbf{x}, \mathbf{z}) + d(\mathbf{z}, \mathbf{y})$

All of the above measures are metrics. Although measures with metric properties are the ones which have been most widely used on interval scale variables others have also been used. Of these non-metric ones the correlation coefficient ρ is perhaps the most widely applied. We are familiar with ρ being used to correlate variables, i.e. as a measure of similarity between variables. In a formally exactly analogous way we can use it to calculate similarities between data points. For example, let x_{ij}, $i = 1, \ldots, n$; $j = 1, \ldots, d$ be the data matrix on n points in d dimensions. For the usual usage the correlation coefficient between variables is estimated by

$$\hat{\rho}_{kj} = \frac{\sum_{i=1}^{n} (x_{ik} - \bar{x}_k)(x_{ij} - \bar{x}_j)}{\sqrt{\sum_{i=1}^{n} (x_{ik} - \bar{x}_k)^2 \sum_{i=1}^{n} (x_{ij} - \bar{x}_j)^2}}$$

where

$$\bar{x}_k = \frac{1}{n} \sum_{i=1}^{n} x_{ik}$$

Now, for correlations between objects we have

$$d_6(\mathbf{x}_k, \mathbf{x}_j) = \hat{\rho}_{kj} = \frac{\sum_{i=1}^{d} (x_{ki} - \bar{\bar{x}}_k)(x_{ji} - \bar{\bar{x}}_j)}{\sqrt{\sum_{i=1}^{d} (x_{ki} - \bar{\bar{x}}_k)^2 \sum_{i=1}^{d} (x_{ji} - \bar{\bar{x}}_j)^2}}$$

with

$$\bar{\bar{x}}_k = \frac{1}{d} \sum_{i=1}^{d} x_{ki}$$

The role of variables and objects is simply reversed. Here, perhaps more obviously than with the metric scales discussed above, the problem associated with using different units becomes apparent: what is the meaning of the mean

$\bar{x}_k$ of measurements on different things? (An average of apples and oranges is what?)

Moving now from interval scales to the opposite extreme of binary variables, many similarity (or association) coefficients have been proposed based on the agreement/disagreement table below (a similarity measure is simply the inverse of a distance measure):

		Object 1	
		1	0
Object 2	1	A	B
	0	C	D

Here both objects simultaneously score 1 on A variables, both simultaneously score 0 on D variables, object 1 scores 1 while object 2 scores 0 on C variables, and object 1 scores 0 while object 2 scores 1 on B variables. Thus, $A + B + C + D = d$, the total number of variables. One of the most common coefficients based on this table is the *simple matching coefficient*

$$d_7(\mathbf{x}, \mathbf{y}) = (A + D)/(A + B + C + D)$$

This is simply the proportion of components in which the two objects agree. Other coefficients omit D from their definitions on the grounds that a match between absent characteristics is not really a match at all (for example, cars, worms, and books do not have legs, but this is hardly a reason for supposing them similar). In this class lie the Jaccard coefficient

$$d_8(\mathbf{x}, \mathbf{y}) = A/(A + B + C)$$

and the Dice coefficient

$$d_9(\mathbf{x}, \mathbf{y}) = 2A/(2A + B + C)$$

The Dice coefficient emphasizes the relative importance of a positive match. In choosing a coefficient caution should be exercised: if code 1 signifies the absence of a characteristic then A rather than D should be omitted and in some cases (e.g. 0 = male, 1 = female) neither code signifies absence.

Because coefficients d_7, d_8, and d_9 are based on matches between components they are called *matching coefficients*. Another type of measure can be derived from the agreement/disagreement table from a probabilistic argument. $A/(A + C)$ is the conditional probability that object 2 will score 1 on a randomly chosen variable given that object 1 scores 1 on that variable. A similar interpretation applies to the probability $A/(A + B)$. A symmetric measure of the strength of relationship between the two objects is thus given by

$$d_{10}(\mathbf{x}, \mathbf{y}) = \frac{1}{2}\left[\frac{A}{A + C} + \frac{A}{A + B}\right]$$

One of the advantages of binary variables is that there are no problems of units of measurement; since they are all recoded to 0 or 1 it does not matter what the original units are. That being said, one can still rescale using weights w_i to make use of any information about the relative importance of the variables. The practical effect is that instead of a match or mismatch on variable i adding 1 to the appropriate cell of the table it now adds w_i.

The principles behind matching coefficients for binary variables can also be applied to nominal variables. Thus, from d_7,

$$d_{11}(\mathbf{x}, \mathbf{y}) = \frac{\text{number of variables on which the objects fall in the same category}}{\text{total number of variables}}$$

Sometimes agreements in certain categories are more important than in others. Thus, if blood group was a variable then the fact that two people both had a rare blood group would be of greater import than if they shared a common blood group. Anderberg (1973) describes a simple and intuitively appealing procedure for weighting agreements: for a particular variable suppose that n_i of the objects fall in category i. Then for two randomly selected objects

$$P\text{ (both fall in category } i) = (n_i/n)^2 \triangleq P_{ii}$$
$$P\text{ (one falls in category } i \text{ and the other in category } j) = 2n_i n_j/n^2 \triangleq P_{ij}$$

and the weights are given by

$$w_{ij} = \frac{1}{P_{ij}} \cdot \frac{2}{g(g+1)}$$

where g is the number of categories for this variable. This simply weights agreement inversely according to their probability of occurrence. The right-hand factor normalizes each variable to unit mean. As for binary variables the weights are used in place of simple counts in matching coefficients.

One can treat ordinal variables as nominal by ignoring the order information, or one can assign numerical ranks to the categories and treat an ordinal variable as interval. We showed above in some detail how the correlation coefficient could be inverted so that instead of measuring similarities between variables it measured similarities between objects. This is a principle which can be applied perfectly generally; so, for example, rank order correlations could be applied to calculating similarities between objects measured on ordinal variables. (But the fact that it can be done does not imply that interpretation is simple.) This also implies a fundamental similarity between cluster analysis and variable selection. We can view variable selection as being a process of choosing representatives from classes of similar variables—so part of the process involves grouping the variables, just as cluster analysis groups the objects. One implication of this is that cluster analysis methods can be applied to variable selection (see, for example, King, 1967 and Mays, 1978).

7.3 HIERARCHICAL METHODS

Hierarchical cluster analysis methods form the final classes by hierarchically grouping subclusters or splitting parent clusters. Thus, one could begin with the n subclusters consisting of one point each and combine these to form larger subclusters then combine these to form still larger ones, and so on until the desired number of clusters has been achieved. Clearly, at each step one would combine those subclusters which were in some sense most similar. Alternatively, one could begin with the single cluster consisting of all the sample points and split this, yielding two clusters. Splitting these will yield more smaller clusters, and so on until again the desired number had been achieved. These two approaches, though hierarchical, are clearly complementary. They are called *agglomerative* and *divisive* methods, respectively. The majority of the published applications of hierarchical cluster analysis techniques seem to make use of agglomerative methods so they will be our chief concern.

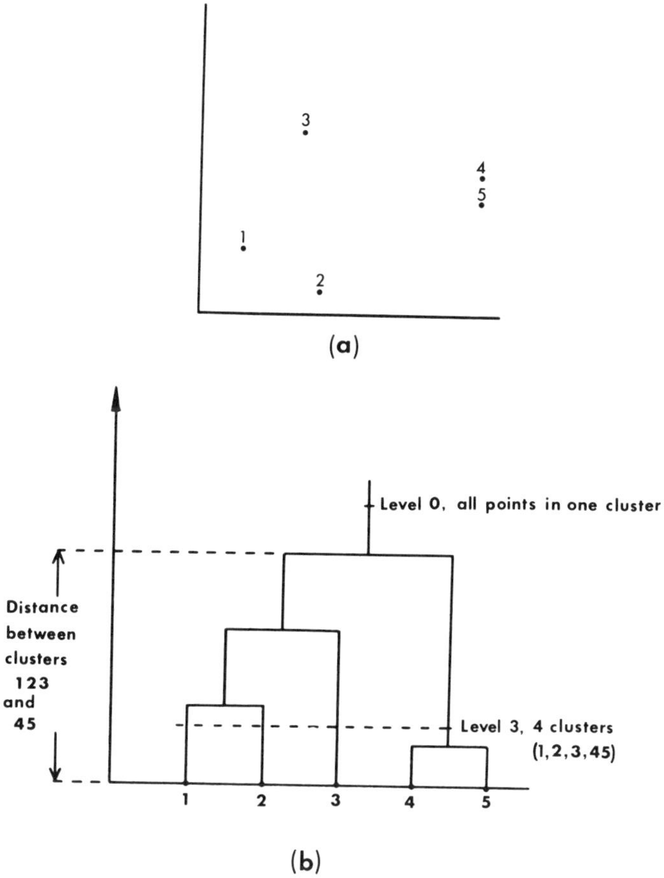

Figure 7.2 Diagram (b) shows the dendrogram for a hierarchical cluster analysis technique applied to the data of diagram (a)

There is a particularly simple way of representing the results of a hierarchical cluster analysis by means of tree graphs or *dendrograms*. Figure 7.2(b) illustrates a dendrogram for a hierarchical techniques applied to the five points shown in Figure 7.2(a). Each level indicates the merging of two subclusters (or splitting, if one is working down the diagram). The vertical axis provides a scale by which to measure the distance or dissimilarity between two merged clusters.

It was remarked above that one of the reasons for the existence of many different cluster analysis techniques is that there are many different ways of defining 'cluster'. For hierarchical techniques these differences are generated by different ways of deciding which pair of clusters should next be merged (in an agglomerative approach) or which cluster should next be split (in a divisive approach). Some of the different ways are presented below.

The chief difference between the methods presented in this section and those of Section 7.4 is that here a point, once assigned to a cluster, cannot be transferred to another (though clearly, in a divisive method, it can go to any of the subclusters generated from its parent cluster). Optimization methods consist of transferring points between clusters with the aim of optimizing some clustering criterion. Although it may seem that this is a major disadvantage of hierarchical methods, it carries the concomitant advantage that hierarchical methods are computationally much quicker than optimization methods. In practice the apparent disadvantage has been found to be not too severe—hierarchical methods usually obtain good solutions when these exist. (Note that in any case the computational requirements of optimization methods mean that often it is necessary to be content with finding a local optimum of the criterion rather than a global one. This is discussed fully in Section 7.4.)

7.3.1 Agglomerative methods

Suppose that an agglomerative method has reached the stage of having c clusters. The next step is to merge two of them into one to yield c-1 clusters. This is then repeated to give c-2 clusters, and so on. Initially, of course, $c = n$, the number of observation points. At each step the two clusters to be merged are chosen by studying the distance (or similarity) matrix of inter-cluster distances (or similarities). For convenience we shall continue the discussion in terms of distances—the interpretation for similarities is straightforward. The natural candidates for merging at each stage are the two clusters which are closest, and it is because there are different ways of measuring inter-cluster distances that there are different agglomerative techniques. However, since all hierarchical methods are fundamentally the same we shall simply list some ways that inter-object distance has been generalized to inter-cluster distance and give a few illustrative examples.

Measure M1. Nearest neighbour

The distance between two clusters is the distance between the closest points, one from each of the two clusters.

This measure is the basis of the *nearest neighbour* or *single link* method.

To illustrate the technique consider Table 7.1 (from Boughey, 1975, p. 338, reproduced with permission). Here we have six objects measured on three variables and we would like to know if the different objects fall into groups of similar objects.

Table 7.1 Forms of intra-uterine devices assessed on three variables

	Events per 100 woman years in use:		
	Pregnancy	Expulsion	Medical removal
A Dalkon Shield	1.1	2.3	2.0
B Gynekoil	1.3	25.8	22.1
C M Device	1.7	2.7	13.5
D Lippes Loop	2.8	10.4	14.0
E Shamrock	4.0	13.0	17.0
F Birnberg Bow	4.7	2.6	14.3

(From *Man and the Environment*, A. S. Boughey, 1975, Macmillan Publishing Co. Inc. Reproduced with permission from the publishers.)

Initially all of the clusters consist of one point each and the inter-cluster distance matrix is simply the inter-point distance matrix. The lower half of this matrix, using squared Euclidean distance, is shown in Table 7.2(a). From this table it is apparent that the two closest clusters are points C and F, so these are merged. The lowest branch of the dendrogram (Figure 7.3) shows this merge.

The new inter-cluster distance matrix is shown in Table 7.2(b). Since we are using the nearest neighbour technique the distance between a point and the cluster (CF) is the distance between the point and the nearest member of (CF). From this table we can see that the next two clusters to be merged are D and E. In computing the new inter-cluster distance matrix, Table 7.2(c), the distance between (DE) and (CF) is the distance between their closest elements, one from each cluster, namely C and D.

This procedure is repeated, producing Tables 7.2(d) and (e) and the complete dendrogram of Figure 7.3. From this dendrogram one might conclude that there are three clusters: one consisting of point *A*, one of point *B*, and one of the remaining points, or perhaps four clusters (A), (B), (DE), and (CF).

One may feel that the raw data should not be used in this analysis because the shorter range of the variable 'pregnancy' means that it does not contribute

Table 7.2 Distance matrix for the data of Table 7.1

	A	B	C	D	E	F
			(a)			
A	0					
B	956.30	0				
C	132.77	607.73	0			
D	212.50	305.02	60.75	0		
E	347.90	197.14	123.63	17.20	0	
F	164.34	610.64	9.65	64.54	115.94	0
			(b)			
	A	B	D	E	(CF)	
A	0					
B	956.30	0				
D	212.50	305.02	0			
E	347.90	197.14	17.20	0		
(CF)	132.77	607.73	60.75	115.94	0	
			(c)			
	A	B	(DE)	(CF)		
A	0					
B	956.30	0				
(DE)	212.50	197.14	0			
(CF)	132.77	607.73	60.75	0		
			(d)			
	A	B	(DECF)			
A	0					
B	956.30	0				
(DECF)	132.77	197.14	0			
			(e)			
		B	(ADECF)			
B		0				
(ADECF)		197.14	0			

enough to the cluster structure. Performing a nearest-neighbour analysis on the standardized data produces the dendrogram of Figure 7.4.

One property of the nearest-neighbour method which is often described as a drawback is that of chaining. This is illustrated in Figure 7.5 where it is seen that a series of points, each one lying near only a few others, has led to some very distant points being grouped into the same cluster.

Measure M2. Furthest neighbour

The distance between two clusters is the distance between the two furthest points, one from each of the two clusters.

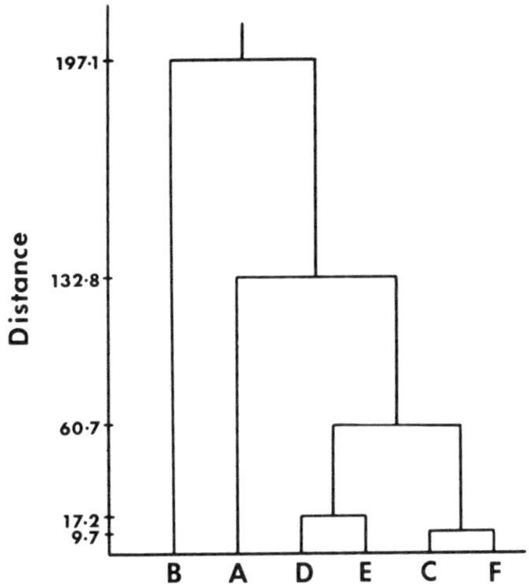

Figure 7.3 The dendrogram for the nearest-neighbour cluster analysis algorithm applied to the data of Table 7.1

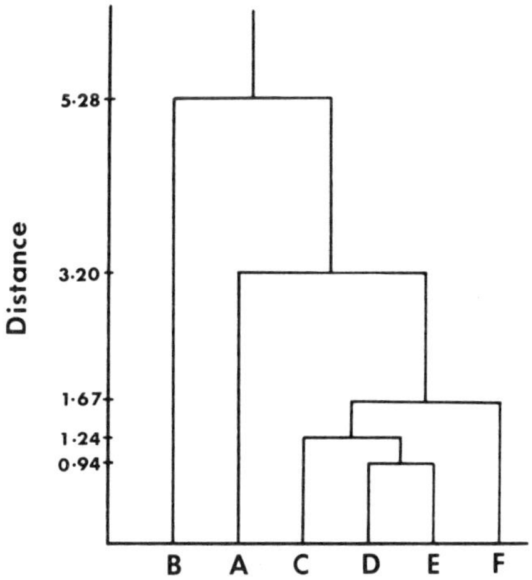

Figure 7.4 The dendrogram for the nearest-neighbour cluster analysis algorithm applied to standardized data from Table 7.1

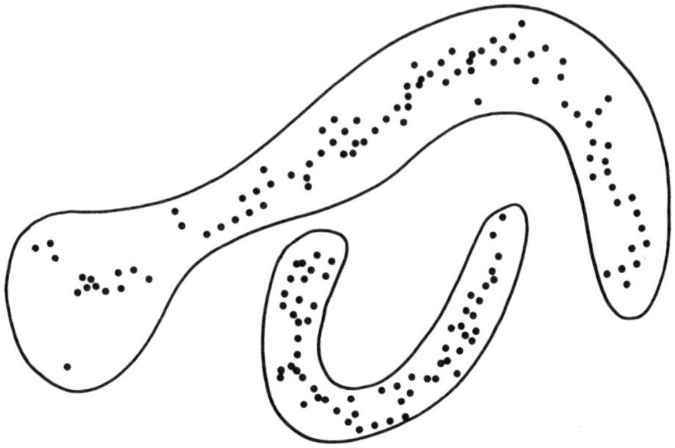

Figure 7.5 Chaining in the nearest-neighbour method

Thus, taking as an example the data of Table 7.1 and the distance matrix of Table 7.2(a), the first merge is C with F. (Since each cluster consists of only one point the 'distance between the two furthest points, one from each cluster' is simply the distance between the two points comprising the two clusters.) For this distance measure, however, the last line of Table 7.2(b) becomes (164.34, 610.64, 64.54, 123.63, 0), with the minimum of the distances A to C and A to F being replaced by the maximum of these distances, etc. Although distance is defined in terms of the most separate points, we still merge the two nearest clusters, of course. The next merge is thus D with E (by chance the same as the nearest neighbour method).

The approach based on measure M2 is called the *furthest neighbour* or *complete link* method.

Measure M3. Centroid distance

The distance between two clusters is the distance between the means (or centroids) of the clusters.

Outliers from even relatively well separated clusters can considerably affect the results produced by measures M1 and M2. Measure M3 attempts to reduce this effect by 'summarizing' the clusters.

Measure M4. Median distance

The distance between two clusters is the distance between the medians of the clusters.

When small clusters are merged with large ones using measure M3, the centroid of the result will lie much nearer the large cluster. Sometimes this can be a disadvantage in that the attributes of the smaller cluster will be largely lost. Measure M4 attempts to overcome this.

It should be realized that the properties of M3 and M4 described above are in a sense complementary and, although these properties are described in terms of 'advantages' and 'disadvantages', whether such labels are appropriate or not will depend on the context of application. This is a general point as far as cluster analysis goes, and one that cannot be over-emphasized: the different methods have different properties and what is a disadvantage for one problem could well be an advantage for another.

Measure M5. Group average distance

The distance between two clusters is the average distance between all pairs of individuals, one taken from each of the two clusters.

Measure M6. Sum of squared deviations

The distance between two clusters is the sum of squares from the objects to the joint cluster mean minus the sum of squares from the objects to their individual cluster means. That is to say, distance is the increase in total sum of squares which would result if the two clusters were combined.

This distance measure is the basis for Ward's (1963) method. The sum of squared distances from the cluster mean is a popular criterion in cluster analysis (indeed, in other guises it is popular throughout statistics) and is also used as a criterion in optimization methods.

Lance and Williams (1967) give a recurrence formula which, with appropriate choice of parameters, enables the above methods to be implemented from a single computer program. This is further discussed in Wishart (1969c).

7.3.2 Divisive methods

Divisive methods begin with the entire set of observations considered as a single cluster and split it into two subclusters. One or the other of these is then split into further subclusters, and so on. The decisions about which cluster to split and how to split it can be based either on variables considered one at a time, or on all variables considered simultaneously. The former are called *monothetic* and the latter *polythetic* techniques.

Examples of monothetic techniques are association analysis (e.g. Lambert and Williams, 1966, and MacNaughton-Smith, 1965) and the automatic interaction detector (Sonquist and Morgan, 1963, 1964). Asssociation analysis takes objects which have been measured on d binary variables, so that each variable imposes an implicit dichotomy on the objects (i.e. each variable divides the set of objects into two classes: those with a score of 0 and those with a score of 1 on the variable). The variables are then searched to find that dichotomy which maximizes some dissimilarity criterion. A common criterion is

$$\sum_{\substack{j=1 \\ j \neq k}}^{d} \chi_{jk}^2, \qquad k = 1, \ldots, d$$

(where χ_{jk}^2 is the chi-squared association coefficient between variables x_j and x_k computed from the 2×2 table of the marginal distribution). The split is made on that k which maximizes this.

The automatic interaction detector approach is similar except that it imposes a binary split on polychotomous variables (by finding that division point which maximizes the criterion) and the criterion is usually the sum of squares in a dependent variable (i.e. the sum of squares of the objects from their respective group means—the unexplained sum of squares). An excellent and brief outline of this sort of approach is given in Fielding (1977).

An example of a polythetic technique is provided by MacNaughton-Smith *et al.* (1964). Objects are gradually transferred from the main cluster to a subcluster by selecting for transfer at each step that object whose dissimilarity from the main cluster, minus its dissimilarity from the subcluster, is a maximum. This transfer stops when all remaining main cluster objects are more similar to the main group than to the subgroup. When this happens the two subgroups are split in the same way. The dissimilarity coefficient suggested by MacNaughton-Smith *et al.* was as follows (for binary variables): let x_{Aj} be the proportion of objects in a group A scoring 1 on variable j and let x_{Bj} be the matching proportion for a group B. Then the dissimilarity between groups A and B is

$$\sum_j \left\{ (x_{Aj} - x_{Bj})^2 \sum_{k \neq j} \chi_{jk}^2 \right\}$$

where the χ_{jk} are calculated for the combined group A + B. The authors tentatively suggest that for quantitiative data one might extend this coefficient to

$$\sum_j \left\{ (\bar{x}_{Aj} - \bar{x}_{Bj})^2 \sum_{k \neq j} r_{jk}^2 \right\}$$

7.4 OPTIMIZATION METHODS

As stated earlier, the principal practical difference between hierarchical and optimization methods is that the latter allow points to be transferred from one cluster to another if that results in an improved clustering. Whether or not an improvement results from such an exchange is determined from the *clustering criterion*. Optimization methods thus try various allocations of points to clusters in an attempt to find that allocation which optimizes the criterion. Different methods have arisen because different criteria and different methods of searching through the space of possible allocations can be used. First, let us consider the criteria.

7.4.1 Optimization criteria

In deciding which criterion to use in an optimization method close attention must be paid to the cluster structures it is desired to detect. The reason for this is simply that different criteria are optimized by different shapes of cluster;

the implication is that extreme care must be taken not to *impose* a non-existent structure on the data. A practical suggestion is that, when exploring data for possible clusters, many different techniques should be tried.

By far the most popular optimization criteria are those based on the matrix identity

$$\mathbf{T} = \mathbf{W} + \mathbf{B}$$

where

$$\mathbf{T} = \sum_{\mathbf{x} \in X^n} (\mathbf{x} - \bar{\mathbf{x}})(\mathbf{x} - \bar{\mathbf{x}})'$$

$$\mathbf{W} = \sum_{i=1}^{c} \sum_{\mathbf{x} \in X_i} (\mathbf{x} - \bar{\mathbf{x}}_i)(\mathbf{x} - \bar{\mathbf{x}}_i)' \tag{1}$$

and

$$\mathbf{B} = \sum_{i=1}^{c} n_i(\bar{\mathbf{x}}_i - \bar{\mathbf{x}})(\bar{\mathbf{x}}_i - \bar{\mathbf{x}})'$$

where

$X^n = \{\mathbf{x}_1, \ldots, \mathbf{x}_n\}$,

$X_i = \{\mathbf{x}_j \mid \mathbf{x}_j \in \text{cluster } i\}$, n_i is the number of points in cluster i,

$\bar{\mathbf{x}}_i = \sum_{\mathbf{x} \in X_i} \mathbf{x}/n_i$, and $\bar{\mathbf{x}} = \sum_{\mathbf{x} \in X^n} \mathbf{x}/n$, the grand mean.

T can be seen to be the scatter matrix describing the overall deviation of the observation points from the grand mean, **W** is the within-class scatter, giving the deviation of the observation points from their cluster means, and **B** is a weighted sum giving the scatter of the cluster means about the grand mean.

The aim of optimization methods based on this identity is, in some sense, to maximize **B** or minimize **W**. However, except for the special case of $d = 1$ when **T**, **B**, and **W** become scalars, in order to make the optimization meaningful it is necessary to summarize the multivariate matrix structure. (And since this summarization may be performed in many ways, even this single matrix identity has led to an abundance of criteria.)

One way to extract a useful univariate index from equation (1), and perhaps the first way which would occur to someone trained in classical statistics, is to use trace **W** (denoted tr **W** below, it is the sum of the diagonal elements of **W**). We have

Criterion C1

$$\begin{aligned}\operatorname{tr} \mathbf{W} &= \operatorname{tr} \sum_{i=1}^{c} \sum_{\mathbf{x} \in X_i} (\mathbf{x} - \bar{\mathbf{x}}_i)(\mathbf{x} - \bar{\mathbf{x}}_i)' \\ &= \sum_{i=1}^{c} \sum_{\mathbf{x} \in X_i} \operatorname{tr}(\mathbf{x} - \bar{\mathbf{x}}_i)(\mathbf{x} - \bar{\mathbf{x}}_i)' \\ &= \sum_{i=1}^{c} \sum_{\mathbf{x} \in X_i} (\mathbf{x} - \bar{\mathbf{x}}_i)'(\mathbf{x} - \bar{\mathbf{x}}_i)\end{aligned}$$

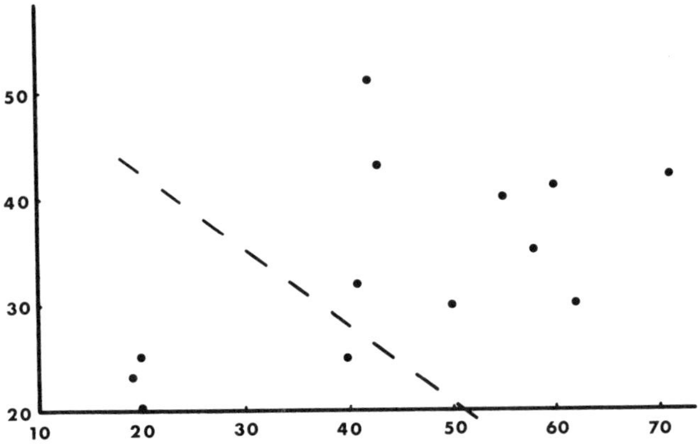

Figure 7.6 A disadvantage (?) of the nearest-neighbour method

Thus, tr **W** is identical to the sum of squared deviations from the observation points to their cluster means.

We have, in fact, already met this criterion in the form of measure M3 during the discussion of agglomerative hierarchical methods. (This leads to another way of looking at hierarchical methods: they can be viewed as being ways of constraining the search through the space consisting of all possible partitions of the objects into clusters. However, they are special ways of constraining the search and do not have the generality of those presented below. In particular, an early poor partition or merging is fixed and cannot be rectified.)

In the context of measures M3 and M4 we noted that outliers or small clusters near large ones could lead to problems. The same arguments apply here and even if the true structure does consist of compact hyperspherical clusters, it could be missed. Figure 7.6 illustrates this, with the line indicating the minimum sum of squares partition.

Aother property (usually a disadvantage?) of this criterion is that tr **W** is not invariant to scaling of the axes so that different results can be obtained on standardized and unstandardized data.

It is in an attempt to overcome this last problem that attention has been turned to alternative criteria which yield solutions which are invariant to linear transformations. A popular criterion in this class is

Criterion C2. $|\mathbf{W}|$, *the determinant of* **W**.

Even though this does not impose the hyperspherical assumption, it does impose the assumption that all clusters have the same shape.

Other invariant criteria have been used on the eigenvalues of the matrix $\mathbf{W}^{-1}\mathbf{B}$ (discussed in Chapter 6). Since the ith eigenvalue, λ_i, is equal to the ratio

(between-cluster scatter)/(within-cluster scatter)

in the direction of the ith eigenvector, the rationale behind such methods is obvious. Also obvious is the fact that at most $(c - 1)$ of the eigenvalues will be non-zero so that criteria such as $\prod_{i=1}^{d} \lambda_i$ are not very useful. Criteria which have been proposed include

Criterion C3

$$\operatorname{tr} \mathbf{W}^{-1}\mathbf{B} = \sum_{i=1}^{d} \lambda_i$$

Criterion C4

$$\operatorname{tr} \mathbf{T}^{-1}\mathbf{W} = \sum_{i=1}^{d} \frac{1}{1 + \lambda_i}$$

Unfortunately (yet another example of all advantages in cluster analysis sometimes being disadvantages), invariance to linear transformations is not as flawless a blessing as it might seem. If different apparent clusterings can be produced by scaling the axes (as in Figure 7.1) then all of these will yield local extrema of linear transformation invariant clustering criteria. The risks will be made more apparent during Section 7.4.2 when the difficulties of avoiding local optima will be pointed out.

Criteria other than those based on partitioning the scatter matrix have been used, though not to any great extent. Interested readers might refer to Rubin (1967) and Wallace and Boulton (1968).

7.4.2 Optimization algorithms

Superficially the problem of finding the optimum (measured by the chosen criterion) partition is straightforward: we simply evaluate the criterion for every possible partition and select the best. Unfortunately, except for very small data sets, there are usually far too many possible partitions and this simplistic approach is totally impossible. Anderberg (1973) gives the number of possible allocations of n objects into c classes as

$$\frac{1}{c!} \sum_{i=0}^{c} (-1)^i \binom{c}{i} (c - i)^n$$

so that, for example, there are 10^{30} possible allocations of 100 objects into 2 classes. It is therefore necessary either to find a more efficient method than exhaustive search or to limit the search to only part of the space of partitions (with consequent risk of missing the optimal one). Many of the methods which have been proposed are based on reformulating the problem so that it becomes suitable for mathematical programming approaches.

Methods based on limiting the scope of the search

Algorithm 1; Evolutionary search

These methods begin with an initial arbitrary clustering as a starting point and consider each object in turn as a candidate for reallocation. If reallocation would result in an improved criterion function value the object is transferred, otherwise it remains in its original cluster. There is, of course, a risk that the search may converge to a local, non-global optimum and to lessen the chance of this it is recommended that the search be repeated from several different starting points.

A popular method based on this sort of principle is the k-means method (MacQueen, 1967). The means of the (k) initial clusters are found and each point is examined to see if it is closer to the mean of another cluster than it is to the mean of its current cluster. If this is the case it is transferred and the cluster means are recalculated. Note that one can either update the means after each point has been reallocated, or else run through all the points and only update when they have all been reallocated. Since updating shifts the means, the points may now no longer be closest to their own cluster means. They are therefore re-examined to see if they are closer to the means of other clusters and the whole process is repeated. Since the mean of a set of objects is that point which minimizes the sum of squares of the distances of the objects to a point it is seen that these methods are based on criterion C1 and that they have a similarity to Ward's hierarchical method (measure M6—the difference being in the method of search through the partition space). McRae (1971) discusses this search method used with other criteria.

To illustrate the technique the basic k-means method (using the tr $\mathbf{W}$ sum of squares criterion) has been applied to the data of Table 7.3 which shows the distribution of rhesus genes in different populations. Note that since the rows each add to 100 per cent one of the columns can be dropped. It is completely determined by the others.

During the first iteration through the data set 8 of the 12 points were reallocated by the ($k = 4$) k-means method. During the second iteration 2 points were reallocated. The final clustering was:

Points 1, 2, and 3 to cluster 1.
Points 4, 5, 7, and 8 to cluster 2.
Points 6, 9, and 10 to cluster 3.
Points 11 and 12 to cluster 4.

It is important to recognize that for these methods the number of clusters (k) must be determined beforehand. There is, however, no reason why a range of k-values should not be tried.

Table 7.3 Distribution of rhesus genes in different populations

Population	Genes							
	CDE	CDe	CdE	Cde	cDE	cdE	cDe	cde
Caucasoid								
1. Danes	0.1	42.2	0.0	1.3	15.1	0.7	1.8	38.8
2. Italians	0.4	47.6	0.3	0.7	10.8	0.7	1.6	38.0
3. Spaniards	0.1	43.2	0.0	1.9	12.0	0.0	3.7	38.0
4. Australian Aborigines (Early Caucasoid)	2.1	56.4	0.0	12.9	20.1	0.0	8.5	0.0
Mongoloid (Recent)								
5. South Chinese	0.5	75.9	0.0	0.0	19.5	0.0	4.1	0.0
6. Japanese	0.4	60.2	0.0	0.0	30.8	3.3	0.0	5.3
Mongoloid (Early)								
7. Eskimos (Greenland)	3.4	72.5	0.0	0.0	22.0	0.0	2.1	0.0
8. Navaho	1.3	43.1	0.0	0.0	27.7	0.0	28.0	0.0
9. Blood	4.1	47.8	0.0	0.0	34.8	3.4	0.0	9.9
10. Chippewa	2.0	33.7	0.0	0.0	53.0	3.2	0.0	8.0
Negroid								
11. Bushmen (Early Negroid)	0.0	9.0	0.0	0.0	2.0	0.0	89.0	0.0
12. Shona (Rhodesia) (Mixed Negroid–Caucasoid)	0.0	6.9	0.0	0.0	6.4	0.0	62.7	23.9

(From *The distribution of the human blood groups*, A. E. Mourant, 1954, Blackwell Scientific Publications Ltd. Reproduced with permission from the publishers.)

Algorithm 2: Random search

Fortier and Solomon (1966) attempt to find the optimum partition by a random search through the partition space. They evaluate the criterion for M randomly generated partitions. Simple probability theory gives the probability that these M will contain one or more from among the best m partitions. In fact, however, they found the results to be disappointing—and attribute this to the highly skewed form of the distribution of the maximum value of the criterion (i.e. the maximum value for each set of M partitions).

In any case from general optimization theory one would not expect such a method to perform as well as more structured search techniques.

Algorithm 3: Steepest descent

Gordon and Henderson (1977) use a hill-climbing (or, more correctly, a hill-descending) approach. They begin by defining a membership matrix **Y** whose

ikth element y_{ik} is 1 if point i belongs to cluster k and 0 otherwise and seek to find that $\mathbf{Y}$ matrix which minimizes

$$S = \sum_{i=1}^{n} \sum_{k=1}^{c} y_{ik} \sum_{j=1}^{d} (x_{ij} - \bar{x}_{ij})^2$$

where x_{ij} is the jth coordinate of the ith point and $\bar{x}_{ij}$ is the jth coordinate of the mean of the cluster to which $\mathbf{x}_i$ belongs. This is simply another way of writing the tr $\mathbf{W}$ sum of squares criterion. By means of a lemma showing that the matrix $\mathbf{Y}$ which minimizes S subject to

$$\sum_{k=1}^{c} y_{ik} = 1, \qquad y_{ik} > 0 \qquad (i = 1, \ldots, n; k = 1, \ldots, c)$$

has elements which are all either 0 or 1 Gordon and Henderson transform the optimization to a more amenable one operating in a continuous $\mathbf{Y}$-space. Further transformations

$$y_{ij} = w_{ij} \Big/ \sum_{k=1}^{c} w_{ik}$$

$$w_{ij} = \exp(v_{ij})$$

change the problem from a constrained to an unconstrained optimization. Now standard iterative steepest descent methods can be applied, resulting in the steps

$$v_{ij}(m+1) = v_{ij}(m) - \lambda \frac{\partial S(m)}{\partial v_{ij}}$$

where λ is a positive scaling factor chosen to give $S(m+1) < S(m)$. The choice of λ is critical, with poor choices leading to poor local minima. The authors recommend imposing an upper bound on λ so that no v_{ij} changes by more than 3 (or 5) at any one step or using a golden section search on λ to find a best value.

In applying the method to an example involving clustering samples of pollen Gordon and Henderson obtained very poor worst runs but best runs which compared favourably with results obtained by other algorithms.

Methods based on efficient complete search

The branch and bound method allows every possible partition to be considered without requiring the explicit evaluation of the criterion function for each partition. It is based on the idea used in Section 2.4 for ordering multivariate data prior to the k-NN pdf estimation method, and in Section 6.4.2 for selecting subsets of variables. The method works with any criterion function which satisfies a certain property. To explain this let us suppose that we want to *minimize* the criterion function. (If, in fact, it is to be maximized, then the inequality below must be reversed.) To set the scene suppose that $A(S_2)$ is an allocation of the points comprising set S_2 into clusters, that $B(S_1)$ is an allocation of the points in S_1 into clusters, that $S_1 \subset S_2$, and that the restriction of A to S_1 is equal to $B(S_1)$. This last simply means that A, when applied only to S_1, gives the same set of clusters as B. Then the property which we require the criterion, J, to satisfy is that

$$J(A(S_2)) \geqslant J(B(S_1))$$

To see how to apply this suppose that we have an allocation, C, of a subset, S_3, of the observation points which gives a greater criterion value than a known allocation, D, of the complete set, S. Then there is no need to investigate allocations of the complete set which include $C(S_3)$. The criterion can only *increase* as we add the extra observation points and already it is greater than $D(S)$, a known allocation of the complete set.

To take a concrete example of how this conceptual principle can be applied in practice suppose we wish to find the optimal partition of four objects (A, B, C, and D) into two clusters (1 and 2). We can enumerate all possible partitions and generate the tree structure shown in Figure 7.7.

In this figure X signifies that a point has not yet been assigned to a cluster and a number i signifies that for this partition the point has been assigned to cluster i. Note that only half of the total tree is shown since the missing half is identical to the illustrated half with 1's replacing 2's and vice versa, i.e. it represents the same clusterings but the clusters have different names.

Now, if we find that

$$J(121X) \geqslant J(1121)$$

there is no point in evaluating partitions 1211 and 1212 since, by virtue of the

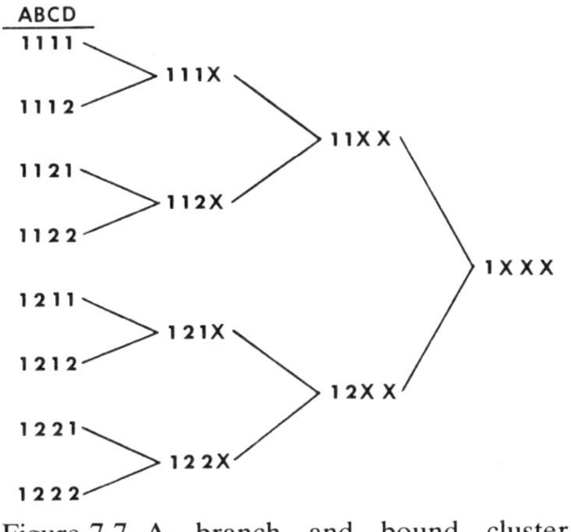

Figure 7.7 A branch and bound cluster analysis tree. 1211 represents the clustering of A, C, and D into one cluster and B into another

property of J defined above, it follows that

$$J(1211) \geqslant J(121X) \geqslant J(1121)$$

and

$$J(1212) \geqslant J(121X) \geqslant J(1121)$$

Similarly, if we find that

$$J(12XX) \geqslant J(1121)$$

we do not need to consider partitions 1211, 1212, 1221, or 1222.

The method thus begins by evaluating criterion J for partitions 1111 and 1112 and selecting the smaller of these. Then $J(112X)$ is evaluated and if

$$J(112X) \geqslant \min(J(1111), J(1112))$$

the search moves to partition $12XX$. Otherwise, partitions 1121 and 1122 are evaluated. Whenever a complete partition with a smaller J-value than any partitions yet discovered is found this J-value acts as an upper bound for all future partitions—no partitions which will have a J larger than this value need be evaluated.

Table 7.4 shows some results obtained by applying this basic branch and bound method to sets of samples from a bivariate normal population with

Table 7.4 Efficiency of the basic branch and bound clustering algorithm for different sample sizes (n) and number of clusters (c). $\bar{N}$ is the number of clusterings tested, $P(n, c)$ is the total number of possible clusterings of n points into c clusters, ρ is the ratio $\bar{N}/P(n, c)$, and ν is the number of simulations carried out for each (n, c) pair

	n			
$c = 2$	5	10	15	20
$\bar{N}$	20	273	2738	55547
$P(n, c)$	15	511	16383	524287
ρ	1.3333	0.5343	0.1671	0.1059
$\max(N)$	28	953	10298	169993
ν	50	50	50	10

	n			
$c = 3$	5	10	15	20
$\bar{N}$	43	1253	16473	573124
$P(n, c)$	25	9330	2375101	580606540
ρ	1.7200	0.1343	0.0069	0.0010
$\max(N)$	63	3781	43322	1624026
ν	50	50	24	10

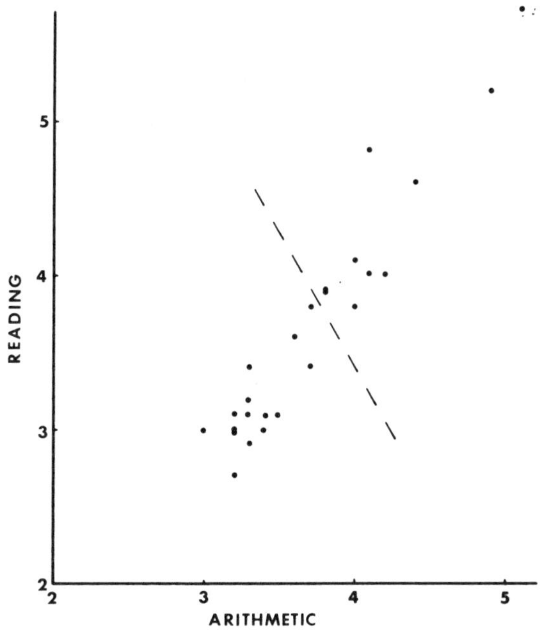

Figure 7.8 The optimum (tr **W**) clustering of the Table 7.5 data obtained by the branch and bound method

identity covariance matrix. Since in such data there is in fact only one cluster the method should perform better (i.e. more quickly) in a real application where there is more structure. ρ is the ratio of the actual number (N) of allocations evaluated to the total number of possible allocations $P(n, c)$. Note that it is possible for ρ to exceed 1 since partial partitions as well as complete partitions are evaluated. The value ν is the number of data sets tested for each n, c pair. $\bar{N}$ is the average number of clusterings evaluated to find the best allocation. In these simulations the tr **W** criterion was used.

Figure 7.8 illustrates the optimum (tr **W**) clustering of the data of Table 7.5 (from Hartigan, 1975) which consists of two scores (the reading and arithmetic averages of fourth-grade children) from 25 schools in America.

While the basic branch and bound method given above extends the range of problems to which an optimum solution can be guaranteed, Koontz *et al.* (1975) have extended the range yet further. They base their improvements on a tighter initial upper bound on J (replacing the smaller of $J(1111)$ and $J(1112)$ by something smaller than both—so that more partitions will be implicitly rejected, without explicit evaluation) and on tighter lower bounds for each partition. Illustrating the latter, in the example above $J(121X)$ would be rejected if it was greater than some value smaller than $J(1121)$. This means that $J(1211)$ and $J(1212)$ have a lower probability of needing explicit evaluation.

Table 7.5 Average reading and arithmetic scores for fourth grade American children from 25 schools

School	Reading	Arithmetic
Baldwin	2.7	3.2
Barnard	3.9	3.8
Beecher	4.8	4.1
Brennan	3.1	3.5
Clinton	3.4	3.7
Conte	3.1	3.4
Davis	4.6	4.4
Day	3.1	3.3
Dwight	3.8	3.7
Edgewood	5.2	4.9
Edwards	3.9	3.8
Hale	4.1	4.0
Hooker	5.7	5.1
Ivy	3.0	3.2
Kimberley	2.9	3.3
Lincoln Bassett	3.4	3.3
Lovell	4.0	4.2
Prince	3.0	3.0
Ross	4.0	4.1
Scranton	3.0	3.2
Sherman	3.6	3.6
Truman	3.1	3.2
West Hills	3.2	3.3
Winchester	3.0	3.4
Woodward	3.8	4.0

The tighter initial upper bound on complete partitions is found by dividing the complete set S into subsets, clustering each of these separately by the basic branch and bound method, and then combining the resulting classifications. Their tighter lower bounds on partitions of subclasses make use of the property that the basic branch and bound method identifies not only the optimal clustering of S, but also the optimal clustering of the first k points of $S (k = 2, \ldots, n)$. Their final improvement is a hierarchical method of combining subsets which leads to yet further improvements while still guaranteeing a globally optimal solution. Readers are referred to Koontz *et al*. (1975) for full details.

Other methods of mathematical programming

Other authors have considered special cases or adapting general cases so that other mathematical programming techniques can be applied.

Lefkovitch (1978), for example, defines measures between sets of objects which allow him to choose a set of subsets such that 'with high probability, the selected ensemble of subsets includes among them the optimal arrangement(s)

of objects', He then formulates the search among the selected ensemble of subsets as a linear programming problem.

Jensen (1968) suggests a dynamic programming method of minimizing

$$H = \sum_{k=1}^{c} T(g_k)$$

where

$$T(g_k) = \frac{1}{n_k} \sum_{i<j \in g_k} d_{ij}^2,$$

d_{ij} is the distance between objects i and j and g_k is the set of objects in the kth cluster. His approach, like the branch and bound method, ensures convergence to the optimal solution without the need to calculate H for all possible clustering alternatives. It is based on the recursion formula

$$H_m^*(z) = \begin{cases} 0, & \text{for } m = 0 \\ \min_y[T(z - y) + H_{m-1}^*(y)], & \text{otherwise} \end{cases}$$

(where $H_m^*(z)$ is the minimum of the criterion for the partition of z objects into m classes, T is the sum of squares of distances within the set of $(z-y)$ objects (as above), and y is a subset of z). For any z the y set which minimizes $H_m^*(z)$ can be found and hence $H_m^*(z)$ can be found for a range of possible z sets. This allows us to find the value of $H_{m+1}^*(x)$, where x is a set including z. And so on until we have the complete set. Interested readers are referred to Jensen (1968) and Bellman (1957).

Although this method guarantees that the optimal solution will be found, Jensen points out that the method requires more computer memory than exhaustive search.

Vinod (1969) uses linear integer programming for the special case of clustering to minimize the within-groups sum of squares for univariate data. Rao (1971) gives the recursive relationships of an efficient dynamic programming approach to the same problem. For the case of multivariate data Rao gives a linear integer programming formulation when the following condition is satisfied: 'in an optimal solution, each group should consists of *at least* one entity, which for convenience will be denoted as the leader of the group, such that the distance between the leader and any entity that does not belong to the same group is not less than the distance between the leading and any entity within the same group.' Although the resulting problem is computationally feasible, it seems to be a harsh constraint, and if it is true one could use methods such as k-means.

Rao also considers minimizing the sum of average within-group squared distances, giving a non-linear integer programming problem. If the number of entities in each group is known, then Rao suggests two simpler approaches: constrained non-linear Boolean programming or linear integer programming.

For minimizing the maximum within-group distance Rao expresses the clustering problem as an integer linear programming exercise, but with a possibly prohibitive number of constraints except for small n and c.

Readers interested in pursuing these approaches are referred to Rao (1971), who also gives references to literature explaining the mathematical programming methods.

7.5 OTHER METHODS

Although the partition into hierarchical and optimization methods is convenient for didactic purposes, there are techniques which do not fit happily into either category. (Indeed, the reader might feel that some of the methods outlined in these sections would fit better elsewhere.) In this section a brief summary is made of some other methods which have been used.

7.5.1 Method of mixtures

In Chapter 3 the estimation of the parameters of mixture distributions was discussed. A similar model can be used for cluster analysis with each mode of pdf, i.e. each component of the mixture, being a cluster. The discussion of Chapter 3 applies directly, with the proviso that the problem of non-identifiability is even more important here. Since our aim is now not simply to describe the overall pdf shape (which is independent of the way it is decomposed) but to separate the objects into natural classes, completely different results will follow from two different decompositions. However, for mixtures of continuous distributions non-identifiability is seldom a problem.

7.5.2 Wishart's Mode method

This method finds an estimate of the local probability density of the parent distribution in the vicinity of each of the object points. Each high density point serves as a cluster centre, and high density points lying close together are merged into one cluster. Originally the k-NN density estimation method described in Chapter 2 was used, but Wishart found that the results varied considerably for different k-values. He therefore replaced the distance to the kth nearest neighbour by the average distance to the $2k$ nearest neighbours in the density estimate. In fact, since absolute magnitudes of pdf are not needed, it is sufficient to use this average distance in place of the pdf estimate (it is monotonically (decreasing) related to the pdf estimate).

Readers interested in details of this technique may refer to Wishart (1969a, 1969b, 1978).

7.5.3 Clumping techniques

There are circumstances (e.g. in the libraries or information retrieval) where it is not necessary that objects should be assigned to only one class but where more useful results are obtained if multiple classifications are permitted. Examples may be found in Williams *et al.* (1969), Stephenson *et al.* (1970), and Ben-Bassat (1980). Methods of cluster analysis which permit this less restricted form are usually known as *clumping* techniques.

A similar result is obtained by Gitman and Levine's (1970) method of detecting fuzzy sets.

7.6 FURTHER READING

Interest in cluster analysis has grown rapidly in the last few years as is dramatically illustrated by a recent survey paper by Blashfield and Aldenderfer (1978). They quote figures demonstrating that the number of publications making use of cluster analysis is rising exponentially. This growth in interest has been accompanied by a matching growth in the number of textbooks and papers attempting to survey, illustrate, and explain the large variety of methods available. Examples of recent papers which compare cluster analysis algorithms are Milligan and Isaac (1980), Bayne, Beauchamp, Begovich, and Kane (1980), and Dubes and Jain (1979). The first of these compares four hierarchical algorithms on constructed data sets and ranks them in order from best to worst: group average, furthest-neighbour, Ward's method, and nearest-neighbour. Bayne *et al.* use Monte Carlo methods to compare thirteen algorithms on the criterion of percentage misclassification. They used six data sets, being different structures of bivariate normal two-class mixtures. They found that the k-means method was the overall best and the nearest-neighbour was one of the overall worst. For readers new to the field of cluster analysis a useful recent paper is that by Dubes and Jain (1979) which 'summarises in a semi-tutorial manner procedures available in the literature for qualitatively evaluating the results of clustering methods without regard for the subject matter of the data'.

Before moving on to the books, there is one other survey paper which should be mentioned, namely Cormack (1971). Despite being nearly ten years old this paper provides a very useful summary of the methods and problems of cluster analysis.

Turning to textbooks, that by Anderberg (1973) seems to be one of the best available. It provides an excellent summary of the conceptual and philosophical problems which will almost certainly be encountered whenever one considers using cluster analysis: types of variables, how to measure association between variables, between data points, and so on. It also provides one of the most comprehensive discussions available on how to use, evaluate, and interpret cluster analysis methods and results. The book concludes with FORTRAN listings of programs for scale conversions, measures of association, cluster analyses themselves, and programs to aid in interpreting clustering results.

Everitt (1980) provides an elementary introduction to, and a useful survey of, the field. Not only does it give numerical illustrations which are small enough to be followed by hand, but it also provides summaries of the latest technical developments and a comprehensive guide to the source literature. A brief section at the end of the book gives details of cluster analysis computer packages, including addresses from which they may be obtained. This book also has the advantage, for those for whom cluster analysis is merely a means rather than a methodological end in itself, that it is only a quarter of the price of some of the other books mentioned here.

Hartigan (1975) is precisely what the title says, namely a collection of cluster analysis algorithms. Although it does not provide the comparative assessments or the general theory (including the difficulties and dangers of cluster analysis) given by other books, it does give FORTRAN listings for each algorithm.

The book by Jardine and Sibson (1971) has a mathematical orientation and is consequently perhaps not to be recommended to the complete novice. Both this and the book by Sokal and Sneath (1963) have an emphasis on biology, with the latter, as the data suggests, being a useful survey of early work in the field, though omitting the important developments of recent years.

Clifford and Stephenson (1975) provide a general introduction to the field from a fairly non-mathematical viewpoint (the qualification implying that since the subject is inherently mathematical it is not possible to avoid mathematics completely). Their orientation is biological and ecological and they include a good outline of the historical development of biological classification. This book would be a good introductory text for biologists but since it does not deal adequately with the full range of available techniques (such as optimization methods) it should be complemented by one of the others.

A good complement to Clifford and Stephenson would be Duran and Odell (1974) who formulate their descriptions in terms of optimizing a criterion function. It is abstract, and fairly mathematical. Its abstract nature means that it is not especially concerned with any particular areas of application, and it does not contain many examples.

Finally, we come to what must surely be the leading book for biologists facing a cluster analysis problem. Sneath and Sokal (1973) originally intended this book to be a revision of their earlier (1963) volume, but it quickly became apparent to the authors that mere revision and updating would not do justice to the amount of work which had been published during the intervening decade. They therefore produced an entirely new volume. Since the authors rank amongst the top experts in the field of biological taxonomy it is not surprising that the book is oriented heavily towards biology. A short chapter at the end illustrates other areas where the techniques may be applied (but the book has, in any case, more than 500 pages). A large portion of the book is devoted to the theoretical biological problems underlying applications of cluster analysis. An excellent bibliography and references which must surely include almost every published work on biological applications of cluster analysis (up to 1973) make it an ideal source book for biologists interested in the field.

As far as cluster analysis packages goes, the package CLUSTAN by Wishart (1978) seems to be the most comprehensive and readily available. The accompanying manual is very clearly written and the programs' outputs are very well annotated.

EXERCISES

7.1 Let $\mathbf{x}_i$, $i = 1, \ldots, n$, be a sample of points, each measured on three variables. Let $\hat{\mathbf{x}}_i$ be the perpendicular projections of these points onto a (two-dimensional) plane which passes through the sample mean. Show that the plane which minimizes

$$\sum_{i=1}^{n} (\mathbf{x}_i - \hat{\mathbf{x}}_i)'(\mathbf{x}_i - \hat{\mathbf{x}}_i)$$

is spanned by the first two principal components.

7.2 In Section 7.2 we saw how the correlation coefficient could be used as a measure of similarity between two vectors $\mathbf{x}$ and $\mathbf{y}$. In some circumstances the angle between two vectors can also be used as a measure of similarity. What is the relationship between the two measures?

7.3 Using a matching coefficient from Section 7.2 and agglomerative cluster analysis methods from Section 7.3.1, investigate the cluster structure of the enuresis data in Table 5.4.

7.4 In Section 3.3.1 we described the maximum likelihood solution to the problem of estimating the parameters of a mixture of normal distributions. Show that the k-means cluster analysis algorithm (Section 7.4.2) yields an approximation to the maximum likelihood solution. In what circumstances might we expect the approximation to be a good one?

7.5 Generate a sample of 200 points from a bivariate normal distribution with identity variance–covariance matrix and apply several different cluster analysis algorithms. Consider the results in the light of the knowledge that there is really only one cluster.

7.6 Show the Jaccard coefficient $A/(A + B + C)$ and the Dice coefficient $2A/(2A + B + C)$, where A, B, and C are defined in Section 7.2, are monotonically related. Is the simple matching coefficient $(A + D)/(A + B + C + D)$ monotonically related to either of the other two?

CHAPTER 8

Miscellaneous Topics

8.1 ASSESSING A CLASSIFIER

When assessing a classifier we have two questions in mind: first, is its absolute performance good enough, and secondly, is its relative performance good enough? The importance of the first question is obvious. Unless the misclassification rate is acceptable for the problem in hand, the classifier will not do. For a particular problem, if the absolute performance is good enough then the relative performance does not matter, but more generally we want to know how different types of classifiers compare: whether one is usually better than another, under what circumstances we should use a particular type, and so on. We mentioned misclassification rate above, and this is the basis we shall use for the assessment, but it should be noted that comparisons between classifiers are implicit in much of the earlier work. First, variable selection are made by comparing the classifiers which arise from the different sets of variables. As was pointed out in Chapter 6, in this situation it is vital to have performance measures which could be evaluated quickly, so the ideal of error rate was sacrificed in favour of simpler functions which had (we hoped) known relationships with error rate. Secondly, estimation of classifier parameters can be thought of as a comparison of the (infinitely many) classifiers arising from varying the parameters. In Chapter 4 we observed that error rate was analytically intractable and replaced it by functions which permitted application of standard optimization techniques.

The simplest and most obvious way of estimating the misclassification rate of a classifier is by seeing how many of the design set points are misclassified. Unfortunately (as remarked in Chapter 1) this leads to an optimistic result—it underestimates the true error rate of the classifier. The reason for this is clear enough: the classifier has been designed to minimize the number of misclassifications (or some related function) of the design set and unless the design set is perfectly representative of the population distribution the classifier will reflect peculiarities of this set which do not exist in the population. A corollary of this is that if the design set is large enough then the method may yield satisfactory results.

The error rate calculated by reclassifying the design set has become known as the *apparent error rate*. We distinguish between this and the *true error rate* of a classifier, which is the expected error rate of this classifier on future samples (from the same population as the design set, of course). McLachlan (1976) has derived the asymptotic bias of the apparent error rate for the case of two multivariate normal populations.

Since the simple method of reclassifying the design set has an optimistic bias we must consider other ways of estimating the true error rate. Moreover, since parts of this book deal with parametric approaches to classifier design, we shall begin with methods based on parametric models—in particular on the assumption of normal class-conditional pdfs. In fact we have already discussed this in Chapter 6 where we outlined analytic evaluation of the true error rate

$$\varepsilon = \int_{\Omega_1} P(\omega_2)p(\mathbf{x}\,|\,\omega_2)\,\mathrm{d}\mathbf{x} + \int_{\Omega_2} P(\omega_1)p(\mathbf{x}\,|\,\omega_1)\,\mathrm{d}\mathbf{x}$$

under the assumption of normal class-conditional pdfs and where it was further assumed that the parameters were known. Even under these improbable assumptions a closed form solution is not possible—although ε does reduce to a univariate normal integral which can be obtained from tables.

Usually, of course, the parameters will be unknown and will need to be estimated from the design set. If this is done then it is easy to see that one is again optimistically biasing the estimate. After all, both the Ω_i regions and the parameters will have been estimated from the same design set, a set which, as before, will have peculiarities not reflected in the population. Hills (1966) has discussed this optimistic bias.

In view of the difficulty of estimating the true error rate attention has been chiefly concentrated on estimating the expected true rate where the expectation is over design set samples of fixed size n_i from class ω_i. Lachenbruch and Mickey (1968) use empirical methods to compare several estimators and recommend, as usually performing well, a method based on an asymptotic expansion of E(true error rate from class ω_i) due to Okamoto (1963) (but see also, Okamoto, 1968). For small values (say, less than 1.0) of the Mahalanobis D^2 $(= (\bar{\mathbf{x}}_1 - \bar{\mathbf{x}}_2)'\mathbf{S}^{-1}(\bar{\mathbf{x}}_1 - \bar{\mathbf{x}}_2))$ between classes, or if the sample size is small relative to the number of parameters, they recommend instead the leaving-one-out or cross-validation method to be outlined below (though of course, if D^2 or the sample is small then caution should be exercised). The leaving-one-out method does not make use of the normality assumption and McLachlan (1974a, 1975a), observing that the Okamoto based method has bias of order 1 (relative to n^{-1}, n^{-1}, N^{-1}, where $N = n_1 + n_2 - 2$) while the leaving-one-out method has bias of order 2, has developed another estimator, based on the normal distribution, with bias of order 3, defined as follows

$$\begin{aligned}\Phi(-D/2) + \phi(-D/2)[\{(d-1)/D\}n_1^{-1} \\ + (D/32)\{4(4d-1) - D^2\}N^{-1} + (d-1)(d-2)(D/4)n_1^{-2} \\ + \{(d-1)/64\}\{-D^3 + 8(2d+1)D + 16/D\}\{n_1 N\}^{-1} \\ + (D/12288)\{3D^6 - 4(24d+7)D^4 \\ + 16(48d^2 - 48d - 53)D^2 + 192(-8d+15)\}N^{-2}]\end{aligned}$$

where ϕ is the standard normal density function, Φ is the standard normal distribution function, and $D = \sqrt{D^2}$, D^2 defined as above.

However, the reader should note that no matter how good a parametric estimator of error rate is, if the true distribution is not that assumed then an inaccurate estimate will result. This is true for parametric models of discrete as well as continuous data. For discrete data, however, it may be possible to use the full multinomial approach—a parameterization which is known to be correct. Hills (1966) considers error rate estimates using the full multinomial approach and based on a normal approximation to the distribution of the difference of probability estimates between classes in each cell. Rather than dwelling on this, however, we shall return to Lachenbruch's (1965) leaving-one-out method which is particularly straightforward to calculate in the case of categorical data. This is also outlined in Hills (1966) and Lachenbruch (1975).

In principle the leaving-one-out method is very simple. One merely classifies in turn each design set point based on the classifier designed on the other $(n-1)$ points. The total number of misclassifications relative to the design set size is then an almost unbiased estimate of the expected true error rate. In general this will clearly involve a lot of computing since n distinct classifiers have to be designed. Frequently, however, the computational burden may be lightened. We will consider normal class-conditional pdfs, kernel pdf estimators, and multinomial estimates.

As we have seen in Chapter 4, for two classes with normal pdfs the optimal discriminant function is

$$f(\mathbf{x}) = D_1^2(\mathbf{x}) - D_2^2(\mathbf{x}) + \ln \frac{|\mathbf{\Sigma}_1|}{|\mathbf{\Sigma}_2|} - 2 \ln \frac{P(\omega_1)}{P(\omega_2)}$$

where $D^2(\mathbf{x}) = (\mathbf{x} - \boldsymbol{\mu}_i)'\mathbf{\Sigma}^{-1}(\mathbf{x} - \boldsymbol{\mu}_i)$, and a common approach is to replace the $\boldsymbol{\mu}_i$, $\mathbf{\Sigma}_i$, and $P(\omega_i)$ by estimates based on the design set, leading to the practical discriminant function $\hat{f}(\mathbf{x})$. Adopting the leaving-one-out method, we must estimate n discriminant functions $\hat{f}_k(\mathbf{x}_k)$ $(k = 1, \ldots, n)$, where the k in f_k indicates that the estimates of the parameters are based on the reduced design set of size $(n-1)$ with $\mathbf{x}_k$ omitted. Fukunaga and Kessell (1971) show that

$$\hat{f}_k(\mathbf{x}_k) = \hat{f}(\mathbf{x}_k) \pm g(n_i, D^2(\mathbf{x}_k)) \tag{1}$$

where the plus sign is taken if $\mathbf{x}_k \in \omega_1$ and the minus sign if $\mathbf{x}_k \in \omega_2$ and where

$$g(n_i, D_i^2(\mathbf{x}_k)) = \frac{(n_i^2 - 3n_i + 1)D_i^2(\mathbf{x}_k)/(n_i - 1) + n_i D_i^4(\mathbf{x}_k)}{(n_i - 1)^2 - n_i D_i^2(\mathbf{x}_k)}$$
$$+ \ln\left[1 - \frac{n_i}{(n_i - 1)^2} D_i^2(\mathbf{x}_k)\right] + 2 \ln \frac{n_i}{n_i - 1}$$
$$+ d \ln \frac{n_i - 1}{n_i - 2} > 0$$

In order to calculate even the straightforward apparent error rate, the first term in (1) must be calculated for each $\mathbf{x}_k$, and this requires $D_1^2(\mathbf{x}_k)$ and $D_2^2(\mathbf{x}_k)$ to be calculated so that there is very little extra work involved in computing $\hat{f}_k(\mathbf{x}_k)$.

If a kernel pdf estimator classifer was being used then in estimating the apparent error rate each design set point, $\mathbf{x}_k$, would be classified on the basis of the pdf estimates at $\mathbf{x}_k$. To apply the leaving-one-out method the only necessary modification is that the pdf estimates are based on the design with $\mathbf{x}_k$ omitted. Thus, no extra computation is required for the leaving-one-out method. Fukunaga and Kessell (1971) demonstrate this in detail for general normal kernels. Similarly, application of the leaving-one-out method to multinomial estimators is also particularly simple—one need only consider those cells for which the classifications change if an observation is removed.

The bootstrap approach to error rate estimation is described by Efron (1979), who shows also that the jackknife method is a linear approximation to this approach. (The bootstrap method is a general and very elegant way of estimating sampling distributions.) Efron's simulations show the bootstrap method outperforming the leaving-one-out method. In terms of mean squared error Efron's bootstrap error rate estimator is

$$(\text{apparent error rate}) - E(B)$$

where B is the bootstrap estimator of the bias in using the apparent error rate to estimate the true rate. B is obtained as follows. For convenience we will consider the two class case and only look at one of the two possible types of misclassification.

Step (i): For class ω_i $(i = 1, 2)$, from the design set (of size n_i) draw with replacement a random sample of size n_i. Call this sample S_i.

Step (ii): Using S_1 and S_2 classify all the n_1 points in the design set for class ω_1. Let the proportion misclassified be m_1.

Step (iii): Using S_1 and S_2 classify the n_1 points in the sample S_1. Let the proportion of these misclassified be $\hat{m}_1$.

Then

$$B = m_1 - \hat{m}_1.$$

Our discussion has been restricted to attempts to design the classifier and estimate the error rate from one set of data. If there is available a separate independent set which can be used as a test set, then application of the classifier to this set yields a straightforward estimate of the error rate. However, if one had such a set available then a better classifier would result if this was combined with the design set so that the classifier was based on a larger set. It has been suggested that one could split the data into two sets; design on one, estimate the error rate from the other, and then recombine them and redesign the classifier from the total set. This has the obvious drawback that the error rate estimated is not that of the classifier finally being used. Compromises between this approach and the leaving-one-out method have also been investigated: for example, dividing the data into ten sets and allocating each of these by the classifier designed on the complementary nine-tenths. An extensive simulation comparison of Lachenbruch's leaving-one-out method with the alternative of splitting the data into independent design and test sets has been made by Penrod and Wagner (1979); see also Devroye and Wagner (1979). Interested readers should also refer to Lachenbruch and Mickey (1968), McLachlan (1974b,c), and Toussaint (1974). The first three of these compare error rate estimators, while the fourth gives a brief description of several such estimators and a bibliography of 188 references on the subject.

Sometimes, instead of straightforward estimates, it may be sufficient to have bounds on the error rate. We have already discussed the Chernoff and Bhattacharyya bounds in Section 6.3. Other recent studies include Boekee and Van der Lubbe (1979), Lapsa (1979), and Devijver (1979). Bounds on the Bayesian error rate provide absolute measures in the sense discussed above and permit one to see if any classifier (based on the chosen variables) can ever be adequate for the purpose in hand.

One final comment about error rate estimation. In Section 8.4 we discuss the use of the 'reject option'. This is simply the possibility of *not* making a classification in all cases (and thus potentially reducing the error rate). By using this idea it is possible to estimate the error rate from an (independent) *unclassified* test set.

8.2 INCOMPLETE DATA

By virtue of their very nature, statistical methods require measurements to be taken from a large number of objects. Because of the large numbers involved it frequently happens that some of the measurements are invalid. They may have been misrecorded, an instrument may have failed, handwriting might be illegible, or they may simply have been lost. Studies extending over time are particularly vulnerable to missing observations—patients drop out of medical studies, subjects move away during sociological investigations, a tractor might destroy a corner plot in an agricultural experiment.

The ideal solution is not to have any missing data—to go back and retake the missing measurements. While this may be practicable sometimes, it frequently happens that it is not. For those situations where it is not possible the most popular solution is to discard any objects with incomplete records. Again this is not always feasible: cases with a large number of measurements on each object arise where virtually none have no missing items. One would end up retaining only a very small fraction of the data set and in any case one would not wish to discard what information does exist in the incomplete vectors. Moreover, discarding a diseased subject as unclassifiable merely because he has lost his legs so that his height cannot be determined would hardly be an acceptable solution. In some problems an additional complication arises in that the mathematics of the analysis possesses certain symmetry properties dependent upon a complete data set and these properties can be destroyed if some data is rejected or is missing.

If rejection of incomplete records is out of the question, what alternative is there? One possibility is to estimate substitute values for the absent items. Note should be taken that this is not an attempt to get something for nothing, it is merely an attempt to retain the simplicity of the existing complete data analysis procedures. If this approach is adopted care must be taken to see that the artificial 'observations' do not contribute spurious information. In the statistical design of experiments a common approach is to estimate substitute values by minimizing the residual sum of squares.

Further information on more general missing data problems may be found in Orchard and Woodbury (1970) and Hoyle (1971). We now turn to the particular problems of missing information in discrimination and classification. Here, as before, we can distinguish two distinct situations: (a) discriminating, i.e. estimating decision rules from incomplete data, and (b) classifying incomplete observation vectors. The second of these problems is the easier and we shall deal with it first (see also Hand, 1978).

One could attempt to estimate substitute values for missing components—an approach which has the merit that it permits the standard complete space classification procedure to be used. It is, however, suboptimal. The optimal solution is to compare the marginal densities of the classes. Suppose, for example, that we have an incomplete observation vector

$$\mathbf{x} = (x_1, ?, ?, x_4, x_5, x_6, ?, x_8)$$

where ? signifies a component is unknown. Then we should compare the class-conditional densities (weighted by priors and costs) at $(x_1, x_4, x_5, x_6, x_8)$ in the five-dimensional subspace spanned by these components. This is equivalent to ignoring the information in the second, third, and seventh components of the design set points.

In general, for the cases treated by Chapters 2 and 3, where the complete space class-conditional pdfs have been estimated, we can obtain the marginal pdfs by appropriate integration. Clearly, however, this could be a lengthy procedure. Fortunately, for many important special cases emphasized in those chapters explicit integration can be avoided.

First consider single multivariate normal pdfs. Thus, let the pdf for one class be

$$f(\mathbf{x}) = \frac{1}{(2\pi)^{d/2}|\mathbf{\Sigma}|^{1/2}} \exp[-\tfrac{1}{2}(\mathbf{x}-\boldsymbol{\mu})'\mathbf{\Sigma}^{-1}(\mathbf{x}-\boldsymbol{\mu})]$$

Now partition $\mathbf{x}$ as $\mathbf{x}' = (\mathbf{x}'_1, \mathbf{x}'_2)$, where $\mathbf{x}_2$ is the vector of unknown values (the result also applies to more general orderings of missing values; it is only expressed in this way for convenience). Partition $\boldsymbol{\mu}$ and $\boldsymbol{\Sigma}$ conformably as

$$\boldsymbol{\mu} = \begin{pmatrix} \boldsymbol{\mu}_1 \\ \boldsymbol{\mu}_2 \end{pmatrix}; \qquad \boldsymbol{\Sigma} = \begin{bmatrix} \boldsymbol{\Sigma}_{11} & \boldsymbol{\Sigma}_{12} \\ \boldsymbol{\Sigma}_{21} & \boldsymbol{\Sigma}_{22} \end{bmatrix}$$

Then the marginal distribution of $\mathbf{x}_1$ is multivariate normal with mean $\boldsymbol{\mu}_1$ and variance–covariance matrix $\boldsymbol{\Sigma}_{11}$. In other words, we can ignore the elements of $\boldsymbol{\mu}$ and the rows and columns of $\boldsymbol{\Sigma}$ which correspond to the missing values. This general theorem applies equally to the components of normal mixtures, of course.

In general, however, since $\boldsymbol{\Sigma}^{-1}$ is not equal to the corresponding quadrant of $\boldsymbol{\Sigma}^{-1}$, each classification of an incomplete vector will involve a matrix inversion as well as evaluation of $|\boldsymbol{\Sigma}_{11}|$. Kittler (1978b) outlines a way to ease the calculations and eliminate explicit matrix inversion at each classification. Suppose for simplicity that there is only one element of $\mathbf{x}$ missing, say x_k. Let $\hat{\boldsymbol{\Sigma}}$ be the variance–covariance matrix in the incomplete space, i.e. $\hat{\boldsymbol{\Sigma}}$ is $\boldsymbol{\Sigma}$ without the kth row and column. Furthermore, let σ_{kk} be the (kk)th element of $\boldsymbol{\Sigma}^{-1}$, $\hat{\boldsymbol{\sigma}}_k$ be the kth column of $\boldsymbol{\Sigma}^{-1}$ with σ_{kk} missing, and $\mathbf{S}$ be $\boldsymbol{\Sigma}^{-1}$ without the kth row and column. Then Kittler shows that

$$|\hat{\boldsymbol{\Sigma}}| = \sigma_{kk} |\boldsymbol{\Sigma}|$$

and

$$\hat{\boldsymbol{\Sigma}}^{-1} = \mathbf{S} - \hat{\boldsymbol{\sigma}}_k \hat{\boldsymbol{\sigma}}'_k / \sigma_{kk}$$

Kittler also shows how this method may be repeated to give an iterative approach when more than one component is missing.

If $\boldsymbol{\Sigma}$ is diagonal, as might be the case when mixtures are used, it being found that an adequate fit is provided by assuming independence *within* each of the mixture components, then inverting the reduced variance–covariance matrix is trivial.

Turning to non-parametric approaches, since kernel pdf estimators are simply mixtures with n components, if normal kernels are used the above results apply directly, especially the comment on diagonal $\boldsymbol{\Sigma}$. In this case (of product kernels) one simply ignores factors corresponding to the missing components

$$\begin{aligned} &\int \cdots \int \frac{1}{nh^d} \sum_{i=1}^{n} \prod_{j=1}^{d} K\left(\frac{x_j - x_{ij}}{h}\right) \mathrm{d}x_{r+1} \ldots \mathrm{d}x_d \\ &= \frac{1}{nh^r} \sum_{i=1}^{n} \left\{ \prod_{j=1}^{r} K\left(\frac{x_j - x_{ij}}{h}\right) \prod_{j=r+1}^{d} \int \frac{1}{h} K\left(\frac{x_j - x_{ij}}{h}\right) \mathrm{d}x_j \right\} \\ &= \frac{1}{nh^r} \sum_{i=1}^{n} \prod_{j=1}^{r} K\left(\frac{x_j - x_{ij}}{h}\right) \end{aligned}$$

Similarly, for nearest-neighbour methods one can simply ignore irrelevant components of the design set elements. (Recall, however, that the nearest-neighbour 'pdf estimator' has an infinite integral—so ignoring components is not equivalent to integration.) Note should also be taken of the fact that optimal kernel and nearest-neighbour spread parameters (h and k, respectively) will not have the same values in marginal subspaces as in the complete space. However, using the complete space values will not usually affect the subspace classification results to any great extent.

For methods which concentrate specifically on decision surfaces, such as the linear methods of Chapter 4 or the condensed nearest-neighbour methods of Chapter 2, rather more complicated procedures must be adopted. These will usually take the form of planning beforehand for the possibility of incomplete vectors and preparing decision surfaces in each subspace. Such methods will be limited by the exponential increase in the number of subspaces as d increases.

As a final cautionary note on classifying incomplete vectors, recall that usually we will have no idea of the subspace misclassification rate. Rather than attempting a classification based on too little information it might be better not to try at all.

We turn now to the first problem mentioned above, that of discriminating, i.e. designing classification rules, from a design set which includes incomplete vectors.

Discarding incomplete vectors carries the risks mentioned above and also the additional risk of introducing considerable bias into the result if the missing values follow some kind of pattern. Admittedly, if the latter is the case some bias will doubtless be present whatever we do, but it could be aggravated by simply ignoring incomplete vectors. In what follows we shall assume that there is no pattern to the incomplete vectors. (Note that the mere existence of missing values could induce a pattern, even if they had originally been randomly distributed. Consider, for example, an analysis based on the weight gain of pigs in litters of given sizes. If, during the course of the experiment, a piglet dies, then the consequent redistribution of the sow's milk could shift the position of the vector of the other piglets' weights.)

Apart from discarding incomplete design set vectors two main types of approach have been considered. The first applies to methods based directly on pdf estimation and is to use the marginal distributions arising from the incomplete vectors to supplement the pdf estimates based on the complete vectors. To illustrate this consider a method based on mean vectors and variance–covariance matrices (such as Fisher's method or a normal mixture method). Then all observations on the first component x_1 can be used in estimating μ_1, the mean of the first component, whether or not $x_2, \ldots, x_d$ are present. Similarly, all present pairs (x_i, x_j) can be used in estimating the correlation between the ith and jth components, whether or not the other components are present. Chan *et al.* (1976) investigated this approach for Fisher's method and found that, in fact, it did not perform as well as a replacement method (see below). The same principle of using marginal information has been applied by Titterington (1977) to the case of the kernel estimator used with binary variables (see Section 5.2.2). In this approach the pdf estimate at point $\mathbf{x}$ is given by

$$p(\mathbf{x}) = \sum_{i=1}^{n} \prod_{j=1}^{d} \lambda_j^{1-|x_j-y_{ij}|}(1 - \lambda_j)^{|x_j-y_{ij}|}$$

where $\{\mathbf{y}_i\}$, $i = 1, \ldots, n$, is the design set. $\lambda_j = \lambda$, estimated as described in Section 5.2.2 if y_{ij} is present, and $\lambda_j = \frac{1}{2}$ if y_{ij} is missing.

The second main type of approach is to estimate substitute values for missing items. This has the advantage that any method of discriminant analysis may be applied once the data have been completed. Chan *et al.* (1976) (see also Chan and Dunn, 1972, 1974) performed Monte Carlo studies to compare several different methods for estimating Fisher's linear discriminant from incomplete data. One of the methods they compared was the 'marginal' method outlined above. Another was to use the subset of complete vectors to calculate regression equations of each variable on the others and then to use these regression equations to estimate substitutes for missing values. The method they recommend as generally best (though not under all circumstances—see Chan *et al.* for details) is a modification of this approach. First replace missing values by the sample means of the variables and then regress each variable on all the others using the completed data. Next, use these regression equations to estimate new substitutes for the missing values and repeat the regression analyses. And so on. This approach has also been suggested by others (for example, Jackson, 1968).

Little (1978) has observed that neither of these methods give consistent estimates of the discriminant function coefficients as n increases with the proportion of incomplete vectors greater than a positive constant. He suggests modified methods which do have this property. One of these is as follows. Calculate substitute values by the first Chan *et al.* method above and then adjust the matrix $\mathbf{A}$ of pooled sum of squares and cross products by adding to a_{jk}, for each vector which has x_j and x_k missing, the residual covariance of x_j and x_k given the variables present in the vector. The new variance–covariance matrix estimate is then the adjusted $\mathbf{A}$ matrix divided by $(n_1 + n_2 - 2)$. Little's second method applies this approach iteratively.

8.3 INCORRECTLY CLASSIFIED DESIGN SETS

With the exception of Chapter 7, where there are no design sets, the assumption has been made throughout this book that the design set elements are correctly classified. Clearly the term 'design set' is meaningless if one has no idea if the design set classifications are correct, but less extreme situations can occur. Here we discuss briefly situations where the design set elements have a non-zero probability of being incorrectly classified. Lachenbruch

(1975) gives the example of studying adverse drug reactions where there is a non-zero probability that a mild symptom, such as a headache, may be due to some cause other than the drug being studied.

Lachenbruch (1966) has investigated this problem for the classical case of two multivariate normal classes with equal covariance matrices. Suppose that a proportion α_1 of the n_1 vectors originally classed (i.e. by the 'teacher') as ω_1 really belong to ω_2 and a proportion α_2 of the n_2 classed as ω_2 really belong to ω_1. Now, as outlined in Chapter 4, the usual classifier is found by substituting in

$$f(\mathbf{x}) = [\mathbf{x} - \tfrac{1}{2}(\boldsymbol{\mu}_1 + \boldsymbol{\mu}_2)]'\boldsymbol{\Sigma}^{-1}(\boldsymbol{\mu}_1 - \boldsymbol{\mu}_2) \gtrless \text{constant}$$

estimates $\bar{\mathbf{x}}_1$, $\bar{\mathbf{x}}_2$, and $\mathbf{S}$ for $\boldsymbol{\mu}_1$, $\boldsymbol{\mu}_2$, and $\boldsymbol{\Sigma}$. However, the misclassifications now mean that, for example, $\bar{\mathbf{x}}_1$ does not estimate $\boldsymbol{\mu}_1$, but instead estimates $(1 - \alpha_1)\boldsymbol{\mu}_1 + \alpha_2\boldsymbol{\mu}_2$. Similar results hold for $\bar{\mathbf{x}}_2$ and $\mathbf{S}$. When the misclassifications are taken into account $f(\mathbf{x})$ becomes

$$f^*(\mathbf{x}) = K\left[f(\mathbf{x}) + \frac{\alpha_1 - \alpha_2}{2}\delta^2\right]$$

where

$$K = (1 - \alpha_1 - \alpha_2)\Big/\left[1 - \frac{c_1 + c_2}{n_1 + n_2} \cdot \delta^2\right]$$

$$c_i = \alpha_i(1 - \alpha_i)/n_i, \qquad i = 1, 2$$

$$\delta^2 = (\boldsymbol{\mu}_1 - \boldsymbol{\mu}_2)'\boldsymbol{\Sigma}^{-1}(\boldsymbol{\mu}_1 - \boldsymbol{\mu}_2)$$

Thus, the discriminant function coefficients based on the partially misclassified design set differ from the ideal coefficients by a constant multiple K, and there is also an added constant term.

The error rate of this new classifier is given by

$$P(f^*(\mathbf{x}) < 0 \mid \mathbf{x} \in \omega_1) = \Phi\left[-\frac{\delta}{2}(1 + \alpha_1 - \alpha_2)\right]$$

and

$$P(f^*(\mathbf{x}) > 0 \mid \mathbf{x} \in \omega_2) = \Phi\left[-\frac{\delta}{2}(1 - \alpha_1 + \alpha_2)\right]$$

where Φ is the standard normal distribution.

Note that if $\alpha_1 = \alpha_2$ then the error rate is unaltered by the incorrect classifications and, furthermore, if α_1 and α_2 are known, then adjustments can be made to $f^*(\mathbf{x})$.

Lachenbruch also performs some small sample simulation experiments and concludes that the large sample results hold to a good approximation. Although these results are based on the assumption that the misclassifications occur randomly in the design set, perhaps rather unrealistic as borderline points are more likely to be misclassified, in Lachenbruch (1974) he found that the true error rates were only affected a little by a non-random pattern of misclassification.

Whereas Lachenbruch uses sampling experiments and simulation to investigate the effect of misclassified design samples on Fisher's linear discriminant function, McLachlan (1972) develops asymptotic expansions for the expected values and variances associated with Fisher's classification rule

$$[\mathbf{x} - \tfrac{1}{2}(\bar{\mathbf{x}}_1 - \bar{\mathbf{x}}_2)]'\mathbf{S}^{-1}(\bar{\mathbf{x}}_1 - \bar{\mathbf{x}}_2) \gtrless 0 \Rightarrow \mathbf{x} \in \begin{cases} \Omega_1 \\ \Omega_2 \end{cases}$$

His conclusion is that Lachenbruch's large sample results hold in most cases.

Shanmugam (1972) has presented a more general sequential Bayesian approach although he assumes that $P(\omega_1) = P(\omega_2)$ and that the misclassification probabilities are known and are equal. Thus, if a point $\mathbf{x}$ belongs to class ω_i it is classified by the teacher as coming from this class with probability

$$P(\Omega_i | \omega_i) = \beta, \qquad i = 1, 2 \tag{1}$$

Thus

$$P(\Omega_i | \omega_j) = 1 - \beta, \qquad i, j = 1, 2; i \neq j$$

He also makes the reasonable assumptions

$$p(\mathbf{x} | \omega_i, \Omega_j) = p(\mathbf{x} | \omega_i)$$
$$p(\mathbf{x}_{k+1} | \omega_i, \boldsymbol{\theta}, \boldsymbol{\Psi}_k, \mathbf{W}_k) = p(\mathbf{x}_{k+1} | \omega_i, \boldsymbol{\theta})$$

where $\mathbf{x}_{k+1}$ is the $(k+1)$th design set point, $\boldsymbol{\theta}$ is the unknown parameter vector, and $\boldsymbol{\Psi}_k = \{\mathbf{x}_1, \ldots, \mathbf{x}_k\}$ has classifications $\mathbf{W}_k = (\Omega_1, \ldots, \Omega_k)$ given by the teacher.

Then Shanmugam shows that $p(\boldsymbol{\theta} | \boldsymbol{\Psi}_k, \mathbf{W}_k)$ may be updated by letting

$$p(\boldsymbol{\theta} | \boldsymbol{\Psi}_{k+1}, \mathbf{W}_{k+1}) = p(\boldsymbol{\theta} | (\mathbf{x}_{k+1}, \omega_i), \boldsymbol{\Psi}_k, \mathbf{W}_k)$$

where ω_i is randomly chosen from ω_1, ω_2 with respective probabilities and

$$p(\omega_1 | \mathbf{x}_{k+1}, \Omega_{k+1}, \boldsymbol{\Psi}_k, \mathbf{W}_k)$$

and

$$P(\omega_2 | \mathbf{x}_{k+1}, \Omega_{k+1}, \boldsymbol{\Psi}_k, \mathbf{W}_k)$$

where Ω_{k+1} is the teacher classification of point $\mathbf{x}_{k+1}$. These selection probabilities can be calculated from

$$P(\omega_i | \mathbf{x}_{k+1}, \Omega_{k+1}, \boldsymbol{\Psi}_k, \mathbf{W}_k) = \frac{p(\mathbf{x}_{k+1} | \omega_i, \boldsymbol{\Psi}_k, \mathbf{W}_k) P(\omega_i | \Omega_{k+1})}{\sum_{j=1,2} p(\mathbf{x}_{k+1} | \omega_j, \boldsymbol{\Psi}_k, \mathbf{W}_k) P(\omega_j | \Omega_{k+1})}$$

and

$$p(\mathbf{x}_{k+1} | \omega_i, \boldsymbol{\Psi}_k, \mathbf{W}_k) = \int p(\mathbf{x}_{k+1} | \omega_i, \boldsymbol{\theta}) p(\boldsymbol{\theta} | \boldsymbol{\Psi}_k, \mathbf{W}_k) \, d\boldsymbol{\theta}$$

with $P(\omega_i | \Omega_{k+1})$ being calculated from (1).

Gimlin (1974) has extended this to the case where $P(\omega_1)$ and $P(\omega_2)$ are also unknown. (He also points out an error in Shanmugam's outlined proof of convergence.)

More generally still, Aitchison and Begg (1976) permit N classes and assume that associated with each element of the design set there is a vector $(\Pi_1, \Pi_2, \ldots, \Pi_N)$ giving the relative plausibilities that the vector belongs to each of the classes $\omega_1, \ldots, \omega_N$. They then extend the predictive discriminant analysis approach (see Section 3.3) to this more general situation.

O'Neill (1978) and Ganesalingam and McLachlan (1978) have studied the extreme problem of a completely unclassified design set. O'Neill concentrates on estimating Fisher's linear discriminant function from such data—a problem clearly related to estimating the parameters of mixture distributions (Chapter 3) and to cluster analysis (Chapter 7). Ganesalingam and McLachlan assume univariate normal classes and study the asymptotic efficiency of the allocation rule based on replacing the parameters in Fisher's linear discriminant function by their maximum likelihood estimates when these are obtained from unclassified data.

Aitchison and Lauder (1979) have studied the related problem of the effect of measurement error on the discriminant analysis.

8.4 THE REJECT OPTION

Generally speaking it is clearly the doubtful classifications which contribute most significantly to the error rate. In some situations it is not vital to classify all the observations and thus in these cases one can decrease the error rate by deciding not to classify such observations. Note that, depending on the problem at hand, one can either discard these doubtful cases or one can collect further information on them so that a classification can be made later. (This principle underlies the methods of sequential variable selection, where more measurements are taken on an object until the probability that it will be correctly classified exceeds some confidence bound.) This is the idea behind the reject option—regions of the space where classifications are doubtful are identified and points falling in these 'rejection' regions are not classified.

The region in which the highest local proportion of misclassifications occur is the region near the decision surface. For the Bayes optimal classifier this is

$$R = \left\{ \mathbf{x} \,|\, 1 - \max_i P(\omega_i|\mathbf{x}) > t \right\}$$

where t is a threshold. The smaller t is, the larger will be the rejection region and the fewer are the points which will be classified. We can express this as a decision rule as

$$\text{if } \max P(\omega_i|\mathbf{x}) \begin{cases} > 1 - t, & \text{then classify } \mathbf{x} \\ < 1 - t, & \text{then reject } \mathbf{x} \end{cases}$$

If we define A as the classification region, the complement of R, so that

$$A = \left\{ \mathbf{x} \,|\, 1 - \max_i P(\omega_i|\mathbf{x}) \leq t \right\}$$

then the unconditional probability of rejecting a point is

$$r(t) = \int_R p(\mathbf{x})\,\mathrm{d}\mathbf{x} \tag{2}$$

$p(\mathbf{x})$ being the overall mixture distribution. Also, the unconditional probability of accepting for classification and then correctly classifying a point is

$$c(t) = \int_A p(\mathbf{x}) \max_i P(\omega_i|\mathbf{x})\,\mathrm{d}\mathbf{x}$$

and the unconditional probability of accepting for classification and then incorrectly classifying a point is

$$\begin{aligned} e(t) &= \int_A p(\mathbf{x})\left[1 - \max_i P(\omega_i|\mathbf{x})\right]\mathrm{d}\mathbf{x} \\ &= 1 - c(t) - r(t) \end{aligned} \tag{3}$$

The chief point to note here is that the error rate and the rejection rate have an inverse relationship.

Fukunaga (1972) then shows that

$$e(t) = -\int_0^t s\,\mathrm{d}r(s) \tag{4}$$

so that if the relationship between t and $r(s)$ is known the error rate $e(t)$ can be calculated. This result can be applied to estimating the misclassification rate of a classifier from a test set of unclassified samples—which can be extremely useful if it is relatively cheap to obtain new measurement vectors but expensive to obtain their teacher classifications. The (classified) design set points are used to determine rejection regions for various values of t and then the unclassified samples are used to obtain estimates of $r(t)$ from (2) for these regions. These estimates can then be used in (4) to give the error rate.

The reader should note that R defined above is in terms of the local proportion of misclassifications. A smaller error rate will result if we define the rejection region R^* to be that region in which the greatest *number* of misclassifications occur. Thus

$$R^* = \left\{\mathbf{x}\,|\,p(\mathbf{x})\left[1 - \max_i P(\omega_i|\mathbf{x})\right] > t\right\}$$

It is obvious from (3) that this will yield a lower error rate. Whether such an approach is practicable or not will depend on the pdfs of the problem.

Brofitt *et al.* (1976) have also studied the reject option. They carried out Monte Carlo investigations to compare three approaches: a method based on assumptions of normality, a tolerance region procedure due to Quesenberry and Gessamen (1968), and a rank procedure (see also Randles *et al.*, 1978). They concluded that 'the rank procedure was the only one that adequately controlled the probabilities of misclassification while maintaining relatively small probabilities of not classifying an observation'. Other studies of the reject option include those of Hellman (1970) and Devijver (1979).

McCarthy (1972) discusses a similar problem.

Further grounds for rejecting a point or deferring a decision have been suggested by Aitchison *et al.* (1977) who propose the use of 'atypicality indices' to indicate the possibility of faulty data. An estimate of the pdf at the location of a point to be classified yields a contour of constant pdf through this point for each class. The atypicality index for a particular class is then simply the probability content of this contour. The less typical a point is of its class the closer the index is to 1.0. If the indices are near 1 for all classes for a point then suspicion arises regarding the quality of the measurements.

EXERCISES

8.1 Write a computer program to implement the leaving-one-out method of error rate estimation for normal classes, as outlined in Section 8.1.

8.2 Write a computer program for an iterative regression method of estimating substitutes for missing observations.

8.3 Using the programs of Exercise 8.1 and 8.2 investigate the relationship between d, n, and the error rate as the proportion of missing design set vector elements increases.

8.4 Discuss ways in which the apparent requirement of the leaving-one-out method for n distinct discriminant analyses can be alleviated for the error correction algorithms of Chapter 4.

8.5 As mentioned before, in some applications speed of classification is very important. Investigate the relative speeds of parametric pdf estimator approaches, non-parametric pdf estimator approaches, and linear decision surface approaches when points to be classified can have missing components.

8.6 Discuss the estimation of class prior probabilities, $P(\omega_i)$, when design set elements have a non-zero probability of being incorrectly classified. What are the advantages to be gained by ignoring points lying close to the decision surface (or using the reject option)?

References

Abend, K. and Harley, T. J. (1969) Comments on the mean accuracy of statistical pattern recognisers. *IEEE Transactions on Information Theory*, **IT-15**, May, 120–121.

Aitchison, J. (1975) Goodness of prediction fit. *Biometrika*, **62**, 547–554.

Aitchison, J. and Aitken, C. G. G. (1976) Multivariate binary discrimination by the kernel method. *Biometrika*, **63**, 413–420.

Aitchison, J. and Begg, C. B. (1976) Statistical diagnosis when basic cases are not classified with certainty. *Biometrika*, **63**, 1–12.

Aitchison, J. and Dunsmore, I. R. (1975) *Statistical prediction analysis*. Cambridge University Press, Cambridge, England.

Aitchison, J. and Lauder, I. J. (1979) Statistical diagnosis from imprecise data. *Biometrika*, **66**, 475–483.

Aitchison, J., Habbema, J. D. F., and Kay, J. W. (1977) A critical comparison of two methods of statistical discrimination. *Applied Statistics*, **26**, 15–25.

Ali, S. M. and Silvey, S. D. (1966) A general class of coefficients of divergence of one distribution from another. *J.R.S.S. (B)*, **28**, 131–142.

Anderberg, M. R. (1973) *Cluster analysis for applications*. Academic Press, New York.

Anderson, G. D. (1969) *A comparison of methods for estimating a probability density function*. Ph.D. thesis, University of Washington.

Anderson, J. A. (1972) Separate sample logistic discrimination. *Biometrika*, **59**, 19–35.

Anderson, J. A. (1975) Quadratic logistic discrimination. *Biometrika*, **62**, 149–154.

Anderson, J. A. and Richardson, S. C. (1979) Logistic discrimination and bias correction in maximum likelihood estimation. *Technometrics*, **21**, 71–78.

Backer, E. and De Shipper, J. A. (1977) On the max-min approach for feature ordering and selection. Proc. Seminar on Pattern Recognition, Liege, November.

Bahadur, R. R. (1961) A representation of the joint distribution of responses to n dichotomous items. In *Studies in Item Analysis and Prediction*, Ed. H. Solomon Stanford University Press, Stanford, Calif.

Bartholomew, D. J. (1959) Note on the measurement and prediction of labour turnover. *J.R.S.S. (A)*, **122**, 232–239.

Bartko, J. J., Strauss, J. S., and Carpenter, W. T. (1971) An evaluation of taxometric techniques for psychiatric data. *Class. Soc. Bull.*, **2**, 1–27.

Bayne, C. K., Beauchamp, J. J., Begovich, C. L., and Kane, V. E. (1980) Monte Carlo comparisons of selected clustering procedures. *Pattern Recognition*, **12**, 51–62.

Beale, E. M. L. (1970) Note on procedures for variable selection in multiple regression. *Technometrics*, **12**, 909–914.

Behboodian, J. (1970) On a mixture of normal distributions, *Biometrika*, **57**, 215–217.

Bellman, R. (1957) *Dynamic programming*. Princeton University Press, Princeton, New Jersey.

Ben-Bassat, M. (1980) Multimembership and multiperspective classification: Introduction, applications, and a Bayesian model. *IEEE Transactions on Systems, Man, and Cybernetics*, **SMC-10**, 331–336.

Berkson, J. T. (1955) Maximum likelihood and minimum χ^2 estimation of the logistic function. *J.A.S.A.*, **50**, 130–162.

Binder, D. A. (1978) Comment on 'Estimating mixtures of normal distributions and switching regressions'. *J.A.S.A.*, **73**, 746–747.

Bishop, Y, M. M., Fienberg, S. E., and Holland, P. W. (1975) *Discrete multivariate analysis: Theory and practice*. MIT Press, Cambridge, Mass.

Blashfield, R. K. and Aldenderfer, M. S. (1978) The literature on cluster analysis. *Multivariate Behavioural Research*, **13**, 271–295.

Bloomfield, P. (1974) Transformations for multivariate binary data. *Biometrics*, **30**, 609–618.

Blum, J. R. (1954) Multidimensional stochastic approximation procedure. *Ann. Math. Stat.*, **25**, 737–744.

Boekee, D. E. and Van der Lubbe, J. C. A. (1979) Some aspects of error bounds in feature selection. *Pattern Recognition*, **11**, 353–360.

Boughey, A. S. (1975) *Man and the Environment*. Collier–Macmillan, London.

Breiman, L., Meisel, W., and Purcell, E. (1977) Variable kernel estimates of multivariate densities, *Technometrics*, **19**, 135–144.

Brofitt, J. D., Randles, R. H., and Hogg, R. V. (1976) Distribution free partial discriminant analysis. *J.A.S.A.*, **71**, 934–939.

Brown, T. A. and Koplowitz, J. (1979) The weighted nearest neighbour rule for class dependent sample sizes. *IEEE Transactions on Information Theory*, **IT-25**, 617–619.

Brunk, H. D. and Pierce, D. A. (1974) Estimation of discrete multivariate densities for computer-aided differential diagnosis of disease. *Biometrika*, **61**, 493–499.

Cacoullos, T. (1966) Estimation of a multivariate density. *Annals of the Institute of Statistical Mathematics*, **18**, 178–189.

Campbell, N. A. (1980) Shrunken estimators in discriminant and canonical variate analysis. *Applied Statistics*, **29**, 5–14.

Čencov, N. N. (1962) Evaluation of an unknown distribution density from observations, *Soviet Math.*, **3**, 1559–1562.

Chan, L. S. and Dunn, O. J. (1972) The treatment of missing values in discriminant analysis—I. The sampling experiment. *J.A.S.A.*, **67**, 473–477.

Chan, L. S. and Dunn, O. J. (1974) A note on the asymptotic aspects of the treatment of missing values in discriminant analysis. *J.A.S.A.*, **69**, 672–673.

Chan, L. S., Gilman, J. A., and Dunn, O. J. (1976) Alternative approaches to missing values in discriminant analysis. *J.A.S.A.*, **71**, 842–844.

Chandra, S. (1977) On the mixtures of probability distributions. *Scand. J. Statist.*, **4**, 105–112.

Chandrasekaran, B. and Harley, T. J. (1969) Comments on the mean accuracy of statistical pattern recognisers. *IEEE Transactions on Information Theory*, **IT-15**, 121–123.

Chernoff, H. (1973) Using faces to represent points in k-dimensional space graphically. *J.A.S.A.*, **68**, 361–368.

Chidananda Gowda, K. and Krishna, G. (1979) The condensed nearest neighbour rule using the concept of mutual nearest neighbourhood. *IEEE Transactions on Information Theory*, **IT-25**, 488–490.

Chien, Y. T. and Fu, K. S. (1967) On Bayesian learning and stochastic approximation. *IEEE Transactions on Systems Science and Cybernetics*, **SSC-3**, 28–38.
Choi, K. and Bulgren, W. G. (1968) An estimation procedure for mixtures of distributions. *J.R.S.S. (B)*, **30**, 444–460.
Clifford, D. H. T. and Stephenson, W. (1975) *An introduction to numerical classification*. Academic Press, New York.
Cochran, W. G. and Hopkins, C. E. (1961) Some classification problems with multivariate data. *Biometrics*, **17**, 10–32.
Cormack, R. M. (1971) A review of classification. *J.R.S.S. (A)*, **134**, 321–367.
Costanza, C. M. and Afifi, A. A. (1979) Comparison of stopping rules in forward stepwise discriminant analysis. *J.A.S.A.*, **74**, 777–785.
Cover, T. M. (1972) A hierarchy of probability density function estimates. In *Frontiers of pattern recognition*, Ed. S. Watanabe. Academic Press, New York, pp. 83–98.
Cover, T. M. (1974) The best two independent measurements are not the two best. *IEEE Transactions on Systems, Man, and Cybernetics*, **SMC-4**, 116–117.
Cox, D. R. (1972) The analysis of multivariate binary data. *Applied Statistics*, **21**, 113–120.
Dantzig, G. B. (1963) *Linear programming and extensions*. Princeton University Press, Princeton.
Davisson, L. D. and Schwartz, S. C. (1970) Analysis of a decision-directed receiver with unknown priors. *IEEE Transactions on Information Theory*, **IT-16**, 270–276.
Day, N. E. (1969) Estimating the components of a mixture of normal distributions. *Biometrika*, **56**, 463–474.
Decell, H. P. and Guseman, L. F. (1979) Linear feature selection with application. *Pattern Recognition*, **11**, 55–63.
De Groot, M. H. (1970) *Optimal statistical decisions*. McGraw-Hill, London.
Devijver, P. A. (1979) New error bounds with the nearest neighbour rule. *IEEE Transactions on Information Theory*, **IT-25**, 749–753.
Devroye, L. P. and Wagner, T. J. (1979) Distribution free inequalities for the deleted and holdout error estimates. *IEEE Transactions on Information Theory*, **IT-25**, 202–207.
Draper, N. R. and Smith, H. (1966) *Applied regression analysis*. Wiley, New York.
Dubes, R. and Jain, A. K. (1979) Validity studies in clustering methodologies. *Pattern Recognition*, **11**, 235–254.
Duda, R. and Hart, P. (1973) *Pattern classification and scene analysis*. Wiley, New York.
Duran, B. S. and Odell, P. L. (1974) *Cluster analysis: A survey*. Springer-Verlag, Berlin.
Dvoretzky, A. (1956) On stochastic approximation. *Proc. 3rd Berkeley Symp. Math. Stat. and Prob.* Univ. of Calif. Press, Berkeley, pp. 39–55.
Efron, B. (1979) Bootstrap methods: another look at the jackknife. *Ann. Stat.*, **7**, 1–26.
Elashoff, J. D., Elashoff, R. M., and Goldman, G. E. (1967) On the choice of variables in classification problems with dichotomous variables. *Biometrika*, **54**, 668–670.
El-Sheikh, T. S. and Wacker, A. G. (1980) Effect of dimensionality and estimation on the performance of Gaussian classifiers. *Pattern Recognition*, **12**, 115–126.
Everitt, B. S. (1977) *The analysis of contingency tables*. Chapman and Hall, London.
Everitt, B. S. (1980) *Cluster analysis*. Heinemann Educational Books, London.
Everitt, B. S. (1981) A test of multivariate normality against the alternative that the distribution is a mixture. *Multivariate Behavioural Research*, to appear in vol. 16.
Everitt, B. S. and Hand, D. J. (1981) *Finite Mixture Distributions*. Chapman and Hall, London.

Fielding, A. (1977) Binary segmentation. The Automatic Interaction Detector and related techniques for exploring data structures. In *Analysis of survey data*, Vol. I, Ed. C. A. O'Muircheartaigh and C. Payne. Wiley, New York.

Fienberg, S. E. (1977) *The analysis of cross-classified data*. MIT Press, Cambridge, Mass.

Fisher, R. A. (1936) The use of multiple measurements in taxonomic problems. *Annals of Eugenics*, **7**, 179–188.

Fortier, J. J. and Solomon, H. (1966) Clustering procedures. In *Multivariate analysis*, Ed. P. R. Krishnaiah. Academic Press, New York, pp. 493–506.

Fryer, M. J. (1977) A review of some nonparametric methods of density estimation. *J. Inst. Maths. Applics*., **20**, 335–354.

Fukunaga, K. (1972) *Introduction to statistical pattern recognition*. Academic Press, London.

Fukunaga, K. and Ando, S. (1977) The optimum nonlinear features for a scatter criterion in discriminant analysis. *IEEE Transactions on Information Theory*, **IT-23**, 453–459.

Fukunaga, K. and Kessell, D. L. (1971) Estimation of classification error. *IEEE Transactions on Computers*, **C-20**, 1521–1527.

Fukunaga, K. and Narendra, P. M. (1975) A branch and bound algorithm for computing *k*-nearest neighbours. *IEEE Transactions on Computers*, **C-24**, 750–753.

Fukunaga, K. and Short, R. D. (1980) A class of feature extraction criteria and its relation to the Bayes risk estimate. *IEEE Transactions on Information Theory*, **IT-26**, 59–65.

Furnival, G. M. (1971) All possible regressions with less computation. *Technometrics*, **13**, 403–408.

Gabriel, K. R. (1968) Simultaneous test procedures in multivariate analysis of variance. *Biometrika*, **55**, 489–504.

Ganesalingam, S. and McLachlan, G. J. (1978) The efficiency of a linear discriminant function based on unclassified initial samples. *Biometrika*, **65**, 658–662.

Gates, G. W. (1972) The reduced nearest neighbour rule. *IEEE Transactions on Information Theory*, **IT-18**, 431.

Geisser, S. (1966) Predictive discrimination, *Proceedings of an International Symposium on Multivariate Analysis, Ohio, 1966*, Ed. P. R. Krishnaiah. Academic Press, New York, pp. 149–165.

Gilbert, E. S. (1968) On discrimination using qualitative variables. *J.A.S.A.*, **63**, 1399–1412.

Gimlin, D. R. (1974) A parametric procedure for imperfectly supervised learning with unknown class probabilities. *IEEE Transactions on Information Theory*, **IT-20**, 661–662.

Gitman, I. and Levine, M. D. (1970) An algorithm for detecting unimodal fuzzy sets and its application as a clustering technique. *IEEE Transactions on Computers*, **C-19**, 583–593.

Goldstein, M. (1975) Comparison of some density estimate classification procedures. *J.A.S.A.*, **70**, 666–669.

Goldstein, M. and Dillon, W. R. (1978) *Discrete discriminant analysis*, Wiley, New York.

Goldstein, M. and Rabinowitz, M. (1975) Selection of variables for the two-group multinomial classification problem. *J.A.S.A.*, **70**, 776–781.

Gordon, A. D. and Henderson, J. T. (1977) An algorithm for Euclidean sum of squares classification, *Biometrics*, **33**, 355–362.

Habbema, J. D. F., Hermans, J., and van den Broek, K. (1974) A stepwise discriminant analysis program using density estimation. In *Compstat. 1974*, Ed. G. Bruckman. Physica–Verlag, Vienna, pp. 101–110.

Habbema, J. D. F., Hermans, J., and Remme, J. (1978) Variable kernel density estimation in discriminant analysis. In *Compstat 1978: Proceedings in computational statistics.* Physica-Verlag, Vienna, pp. 178–185.

Hand, D. J. (1978) A survey of techniques for classifying incomplete vectors. In *Data structure analysis and its applications—ENST-C-78002*, Ed. J. Kittler. Ecole Nationale Superieure des Telecommunications, Paris, France.

Hand, D. J. (1981) Branch and bound in statistical data analysis. *The Statistician*, **30**, 1–13.

Hand, D. J. and Batchelor, B. G. (1978) An edited condensed nearest neighbour rule. *Information Sciences*. **14**, 171–180.

Hart, P. E. (1968) The condensed nearest neighbour rule. *IEEE Transactions on Information Theory*, **IT-14**, 515–516.

Hartigan, J. A. (1975) *Clustering algorithms*, Wiley, New York.

Hasselblad, V. (1966) Estimation of parameters for a mixture of normal distributions. *Technometrics*, **8**, 431–444.

Hellman, M. E. (1970) The nearest neighbour classification rule with a reject option. *IEEE Transactions on Systems Science and Cybernetics*, **SSC-6**, 179–185.

Hills, M. (1966) Allocation rules and their error rates. *J.R.S.S. (B)*, **28**, 1–20.

Hills, M. (1967) Discrimination and allocation with discrete data. *Applied Statistics*, **16**, 237–250.

Ho, Y-C. and Kashyap, R. L. (1965) An algorithm for linear inequalities and its applications. *IEEE Transactions on Electronic Computers,* **EC-14**, 683–688.

Ho, Y-C. and Kashyap, R. L. (1966) A class of iterative procedures for linear inequalities. *J. SIAM Control*, **4**, 112–115.

Hora, S. C. (1978) Sample size determination in Bayesian discriminant analysis. *J.A.S.A.*, **73**, 569–572.

Hoyle, M. H. (1971) Spoilt data—an introduction and bibliography. *J.R.S.S. (A)*, **34**, 429–439.

Hughes, G. F. (1968) On the accuracy of statistical pattern recognisers. *IEEE Transactions on Information Theory*, **IT-14**, 55–63.

Jackson, E. C. (1968) Missing values in linear multiple discriminant analysis. *Biometrics*, **24**, 835–844.

Jain, A. K. and Dubes, R. (1978) Feature definition in pattern recognition with small sample size. *Pattern Recognition*, **10**, 85–97.

Jain, A. K. and Waller, W. G. (1978) On the optimal number of features in the classification of multivariate Gaussian data. *Pattern Recognition*, **10**, 365–374.

Jardine, N. and Sibson, R. (1971) *Mathematical taxonomy.* Wiley, New York.

Jaynes, E. T. (1968) Prior probabilities, *IEEE Transactions on Systems Science and Cybernetics*, **SSC-4**, 227–241.

Jensen, R. E. (1968) A dynamic programming algorithm for cluster analysis. *Op. Res.*, 1034–1056.

Kabir, A.B.M.L. (1968) Estimation of parameters of a finite mixture of distributions. *J.R.S.S. (B)*, **30**, 472–482.

Kailath, T. (1967) The divergence and Bhattacharayya distance measures in signal selection. *IEEE Transactions on Computers*, **COM-15**, 52–60.

Keehn, D. G. (1965) A note on learning for Gaussian properties. *IEEE Transactions on Information Theory*, **IT-11**, 126–132.

Kendall, M. G. and Stuart, A. (1961) *The advanced theory of statistics*. Griffin, London.

Kenward, M. G. (1979) An intuitive approach to the MANOVA test criteria. *The Statistician*, **28**, 193–198.

Kiefer, J. and Wolfowitz, J. (1952) Stochastic estimation of the maximum of a regression function. *Ann. Math. Stat.*, **23**, 462–466.

King, B. F. (1967) Stepwise clustering procedures. *J.A.S.A.*, **62**, 86–101.

Kittler, J. (1975) Mathematical methods of feature selection in pattern recognition. *Int. J. Man–Machine Studies*, **7**, 609–637.
Kittler, J. (1978a) Feature set search algorithms. In *Pattern recognition and signal processing*, Ed. C. H. Chen. Publ. Sijthoff and Noordhoff, The Netherlands.
Kittler, J. (1978b) Multiclass parametric decision-making processor for classification of patterns with missing descriptors. *IEE Journal on Computers and Digital Techniques*, **1**, 53–59.
Koontz, W. L. G., Narendra, P. M., and Fukunaga, K. (1975) A branch and bound clustering algorithm. *IEEE Transactions on Computers*, **C-24**, 908–915.
Kronmal, R. and Tarter, M. (1968). The estimation of probability densities and cumulatives by Fourier series methods. *J.A.S.A.*, **63**, 925–952.
Kruskal, J. B. (1964) Multidimensional scaling by optimising goodness of fit to a nonmetric hypothesis. *Psychometrika*, **29**, 1–27.
Krzanowski, W. J. (1975) Discrimination and classification using both binary and continuous variables. *J.A.S.A.*, **70**, 782–790.
Krzanowski, W. J. (1977) The performance of Fisher's linear discriminant function under non-optimal conditions. *Technometrics*, **19**, 191–199.
Krzanowski, W. J. (1979) Some linear transformations for mixtures of binary and continuous variables, with particular reference to linear discriminant analysis. *Biometrika*, **66**, 33–39.
Kullback, S. (1968) *Information theory and statistics*. Dover Publications, New York.
Lachenbruch, P. A. (1965) *Estimation of error rates in discriminant analysis*. Ph.D. dissertation, University of California at Los Angeles.
Lachenbruch, P. A. (1966) Discriminant analysis when the initial samples are misclassified. *Technometrics*, **8**, 657–662.
Lachenbruch, P. A. (1974) Discriminant analysis where the initial samples are misclassified. II: Non-random misclassification models. *Technometrics*, **16**, 419–424.
Lachenbruch, P. A. (1975) *Discriminant analysis*. Hafner Press, New York.
Lachenbruch, P. A. and Mickey, M. R. (1968) Estimation of error rates in discriminant analysis. *Technometrics*, **10**, 1–11.
Lambert, J. M. and Williams, W. T. (1966) Multivariate methods in plant ecology IV: Comparison of information analysis and association analysis. *J. Ecol.*, **54**, 635–664.
Lance, G. N. and Williams, W. T. (1967) A general theory of classificatory sorting strategies: 1. Hierarchical systems. *Computer Journal*, **9**, 373–390.
Lapsa, P. M. (1979) Some statistical bounds for the accuracy of distance-based pattern classification. *Pattern Recognition*, **11**, 95–108.
Laurie, P. (1979) *Beneath the city streets*. Granada Publishing, London.
Lefkovitch, L. P. (1978) Cluster generation and grouping using mathematical programming. *Mathematical Biosciences*, **41**, 91–110.
Liddell, D. (1977) Multivariate response in more than one sample. *The Statistician*, **26**, 1–15.
Lindley, D. V. (1965) *Introduction to probability and statistics*. Cambridge University Press.
Little, R. J. A. (1978) Consistent regression methods for discriminant analysis with incomplete data. *J.A.S.A*, **73**, 319–322.
Loftsgaarden, D. O. and Quesenberry, C. P. (1965) A nonparametric estimate of a multivariate density function. *Ann. Math. Stat.*, **36**, 1049–1051.
McCabe, G. P. (1975) Computations for variable selection in discriminant analysis. *Technometrics*, **17**, 103–109.
McCarthy, P. J. (1972) The effects of discarding inliars when binomial data are subject to classification errors. *J.A.S.A.*, **67**, 515–529.
McCulloch, W. S. and Pitts, W. H. (1943) A logical calculus of the ideas immanent in nervous activity, *Bulletin of Math. Biophysics*, **5**, 115–133.

McKay, R. J. (1976) Simultaneous procedures in discriminant analysis involving two groups. *Technometrics*, **18**, 47–53.

McKay, R. J. (1977) Simultaneous procedures for variable selection in multiple discriminant analysis. *Biometrika*, **64**, 283–290.

McKay, R. J. (1978) A graphical aid to selection of variables in two-group discriminant analysis. *Applied Statistics*, **27**, 259–263.

McLachlan, G. J. (1972) Asymptotic results for discriminant analysis when the initial samples are misclassified. *Technometrics*, **14**, 415–422.

McLachlan, G. J. (1974a) An asymptotic unbiased technique for estimating the error rates in discriminant analysis. *Biometrics*, **30**, 239–249.

McLachlan, G. J. (1974b) Estimation of the errors of misclassification on the criterion of asymptotic mean square error. *Technometrics*, **16**, 255–260.

McLachlan, G. J. (1974c) The relationship in terms of asymptotic mean square error between the separate problems of estimating each of the three types of error rate of the linear discriminant function. *Technometrics*, **16**, 569–575.

McLachlan, G. J. (1975a) Iterative reclassification procedure for constructing an asymptotically optimal rule of allocation in discriminant analysis. *J.A.S.A.*, **70**, 365–369.

McLachlan, G. J. (1975b) Confidence intervals for the conditional probability of misallocation in discriminant analysis. *Biometrics*, **31**, 161–167.

McLachlan, G. J. (1976) The bias of the apparent error rate in discriminant analysis. *Biometrika*, **63**, 239–244.

MacNaughton-Smith, P. (1965) Some statistical and other numerical techniques for classifying individuals. Home Office Research Unit Report No. 6. HMSO, London.

MacNaughton-Smith, P., Williams, W. T., Dale, M. B., and Mockett, L. G. (1964) Dissimilarity analysis. *Nature*, **202**, 1034–1035.

MacQueen, J. (1967) Some methods for classification and analysis of multivariate observations. *Proc. 5th Berkeley Symposium*, **1**, 281–297.

McRae, D. J. (1971) MICKA, a FORTRAN IV iterative k-means cluster analysis program. *Behavioural Science*, **16**, 423–424.

Margolese, M. S. (1970) Homosexuality: A new endocrine correlate. *Horm. and Behav.*, **1**, 151–155.

Marriott, F. H. C. (1971) Practical problems in a method of cluster analysis. *Biometrics*, **27**, 501–544.

Martin, D. C. and Bradley, R. A. (1972) Probability models estimation and classification for multivariate dichotomous populations. *Biometrics*, **28**, 203–221.

Mays, R. (1978) Interactive maximum reliability cluster analysis. *Educational and Psychological Measurement*, **38**, 783–785.

Meisel, W. S. (1972) *Computer oriented approaches to pattern recognition.* Academic Press, New York.

Milligan, G. W. and Isaac, P. D. (1980) The validation of four ultrametric clustering algorithms. *Pattern Recognition*, **12**, 41–50.

Moore, D. H. (1973) Evaluation of five discrimination procedures for binary variables. *J.A.S.A.*, **68**, 399–404.

Moran, M. A. and Murphy, B. J. (1979) A closer look at two alternative methods of statistical discrimination. *Applied Statistics*, **28**, 223–232.

Morrison, D. F. (1967) *Multivariate statistical methods.* McGraw-Hill, London

Mourant, A. E. (1954) *The distribution of the human blood groups.* Blackwell, Oxford.

Mucciardi, A. N. and Gose, E. E. (1972) An automatic clustering algorithm and its properties in high-dimensional spaces. *IEEE Transactions on Systems, Man and Cybernetics*, **SMC-2**, 247–254.

Murray, G. D. (1977) A cautionary note on selection of variables in discriminant analysis. *Applied Statistics*, **26**, 246–250.

Nilsson, N. J. (1965) *Learning machines. Foundations of trainable pattern classifying systems.* McGraw-Hill, New York.

Okamoto, M. (1963) An asymptotic expansion for the distribution of the linear discriminant function. *Ann. Math. Stat.*, **34**, 1286–1301.

Okamoto, M. (1968) Correction to: An asymptotic expansion for the distribution of the linear discriminant function. *Ann. Math. Stat.*, **39**, 1358–1380.

O'Neill, T. J. (1978) Normal discrimination with unclassified observations. *J.A.S.A*, **73**, 821–826.

Orchard, T. and Woodbury, M. A. (1970) A missing information principle: Theory and applications. *Proc. 6th Berkeley Symposium*, **1**, 697–715.

Ott, J. and Kronmal, R. A. (1976) Some classification procedures for multivariate binary data using orthogonal functions. *J.A.S.A.*, **71**, 391–399.

Parzen, E. (1962) On estimation of a probability density function and mode. *Ann. Math. Stat.*, **33**, 1065–1076.

Patrick, E. A. and Fisher, F. P. (1969) Nonparametric feature selection. *IEEE Transactions on Information Theory*, **IT-15**, 577–584.

Pearson, K. (1894) Contributions to the mathematical theory of evolution. *Phil. Trans. (A)*, **185**, 71–110.

Penrod, C. S. and Wagner, T. J. (1979) Risk estimation for nonparametric discrimination and estimation rules: A simulation study. *IEEE Transactions on Information Theory*, **IT-25**, 753–758.

Quandt, R. E. and Ramsey, J. B. (1978) Estimating mixtures of normal distributions and switching regressions. *J.A.S.A.*, **73**, 730–738.

Quesenberry, C. P. and Gessamen, M. P. (1968) Nonparametric discrimination using tolerance regions. *Ann. Math. Stat.*, **39**, 664–673.

Raatgever, J. W. and Duin, R. P. W. (1978) On the variable kernel model for multivariate nonparametric density estimation. In *COMPSTAT 1978: Proceedings in computational statistics*. Physica-Verlag, Vienna, pp. 524–533.

Raiffa, H. and Schlaifer, R. (1961) *Applied statistical decision theory*. Harvard University.

Randles, R. H., Broffitt, J. D., Ramberg, J. S., and Hogg, R. V. (1978) Discriminant analysis based on ranks. *J.A.S.A.*, **73**, 379–384.

Rao, C. R. (1970) Inference on discriminant function coefficients. In *Essays in probability and statistics*, Ed. R. C. Bose *et al*. The University of North Carolina Press, Chapel Hill, pp. 587–602.

Rao, M. R. (1971) Cluster analysis and mathematical programming. *J.A.S.A.*, **66**, 622–626.

Rennie, R. R. (1972) On the interdependence of the identifiability of finite multivariate mixtures and the identifiability of the marginal mixtures. *Sankhya* **34**, 449–452.

Robbins, H. and Monro, S. (1951) A stochastic approximation method. *Ann. Math. Stat.*, **22**, 400–407.

Rosenblatt, F. (1962) *Principles of neurodynamics: Perceptrons and the theory of brain mechanisms*. Spartan Books, Washington, D.C.

Rosenblatt, M. (1956) Remarks on some nonparametric estimates of a density function. *Ann. Math. Stat.*, **27**, 832–835.

Roy, J. (1958) Step down procedure in multivariate analysis. *Ann. Math. Stat.*, **29**, 1177–1187.

Rubin, J. (1967) Optimal classification into groups: An approach for solving the taxonomy problem. *J. Theor. Biol.*, **15**, 103–144.

Sacks, J. (1958) Asymptotic distribution of stochastic approximation procedures. *Ann. Math. Stat.*, **29**, 373–386.

Schwartz, S. C. (1967) Estimation of a probability density by an orthogonal series. *Ann. Math. Stat.*, **38**, 1261–1265.

Sebestyen, G. and Edie, J. (1966) An algorithm for non-parametric pattern recognition. *IEEE Transactions on Electronic Computers*, **EC-15**, 908–915.
Shanmugam, K. (1972) A parametric procedure for learning with an imperfect teacher. *IEEE Transactions on Information Theory*, **IT-18**, 300–302.
Shanmugam, K. (1977) On a modified form of Parzen estimator for nonparametric pattern recognition. *Pattern Recognition*, **9**, 167–170.
Shepard, R. N. (1974) Representation of structure in similarity data: Problems and prospects. *Psychometrika*, **39**, 373–421.
Shepard, R. N., Romney, A. K., and Nerlove, S. B. (1972) *Multidimensional scaling*. Seminar Press, New York.
Sneath, P. H. A. and Sokal, R. R. (1973) *Numerical taxonomy*. W. H. Freeman and Co., San Francisco.
Sokal, R. R. and Sneath, P. H. A. (1963) *Principles of numerical taxonomy*. W. H. Freeman, London.
Sonquist, J. A. and Morgan, J. N. (1963) Problems in the analysis of survey data and a proposal. *J.A.S.A.*, **58**, 415–435.
Sonquist, J. A. and Morgan, J. N. (1964) *The determination of interaction effects*. Survey Research Centre, Institute of Social Research, University of Michigan.
Spragins, J. (1965) A note on the iterative applications of Bayes' rule. *IEEE Transactions on Information Theory*, **IT-11**, 544–549.
Stephenson, W., Williams, W. T., and Lance, G. N. (1970) The macrobenthos of Moreton Bay. *Ecol. Monogr.*, **40**, 459–494.
Tarter, M. E., Holcomb, R. L., and Kronmal, R. A. (1967) A description of new computer methods for estimating the population density. *Proc. A.C.M.*, **22**, 511–519.
Teicher, H. (1960) On the mixture of distributions. *Ann. Math. Stat.*, **31**, 55–73.
Teicher, H. (1961) Identifiability of mixtures. *Ann. Math. Stat.*, **32**, 244–248.
Teicher, H. (1963) Identifiability of finite mixtures. *Ann. Math. Stat.*, **34**, 1265–1269.
Teicher, H. (1967) Identifiability of mixtures of product measures. *Ann. Math. Stat.*, **38**, 1300–1302.
Thorndike, R. L. (1953) Who belongs in the family? *Psychometrika*, **18**, 267–276.
Titterington, D. M. (1977) Analysis of incomplete multivariate binary data by the kernel method. *Biometrika*, **64**, 455–460.
Tou, J. T. and Heydorn, R. P. (1967) Some approaches to optimum feature extraction. In *Computers and Information Sciences*, Vol. II, Ed. J. T. Tou. Academic Press, New York, pp. 57–89.
Toussaint, G. T. (1974) Bibliography on estimation of misclassification. *IEEE Transactions on Information Theory*, **IT-20**, 472–479.
Vajda, S. (1970) *Theory of games and linear programming*. Methuen, London.
Van Campenhout, J. M. (1978) On the peaking of the Hughes mean recognition accuracy: The resolution of an apparent paradox. *IEEE Transactions on Systems, Man, and Cybernetics*, **SMC-8**, 390–395.
Van Der Waerden, B. L. (1969) *Mathematical statistics*. George Allen & Unwin, London.
Van Ness, J. W. (1979) On the effects of dimension in discriminant analysis for unequal covariance populations. *Technometrics*, **21**, 119–127.
Van Ness, J. W. and Simpson, C. (1976) On the effects of dimension in discriminant analysis, *Technometrics*, **18**, 175–187.
Vinod, H. D. (1969) Integer programming and the theory of grouping. *J.A.S.A.*, **64**, 506–519.
Wagner, T. J. (1975) Nonparametric estimates of probability densities. *IEEE Transactions on Information Theory*, **IT-21**, 438–440.
Wahba, G. (1971) A polynomial algorithm for density estimation. *Ann. Math. Stat.*, **42**, 1870–1886.

Wahl, P. W. and Kronmal, R. A. (1977) Discriminant functions when covariances are unequal and sample sizes are moderate. *Biometrics*, **33**, 479–484.
Wallace, C. S. and Boulton, D. M. (1968) An information measure for classification. *Computer Journal*, **11**, 185–194.
Waller, W. G. and Jain, A. K. (1978) On the monotonicity of the performance of a Bayesian classifier. *IEEE Transactions on Information Theory*, **IT-24**, 392–394.
Ward, J. H. (1963) Hierarchical grouping to optimise an objective function. *J.A.S.A.*, **58**, 236–244.
Watanabe, S., Lambert, P. F., Kulikowski, C. A., Buxton, J. L., and Walker, R. (1967) Evaluation and selection of variables in pattern recognition. In *Computers and Information Sciences*, Vol. II, Ed. J. T. Tou. Academic Press, New York, pp. 91–122.
Watson, G. S. (1969) Density estimation by orthogonal series. *Ann. Math. Stat.*, **40**, 1496–1498.
Wegman, E. J. (1972) Nonparametric probability density estimation I: A summary of available methods. *Technometrics*, **14**, 533–546.
Wertz, W. (1978) *Statistical density estimation: A survey*. Göttingen, Vandenhoek, and Ruprecht, Monographs in Applied Statistics and Econometrics, No. 13.
Whittle, P. (1958) On smoothing of probability density functions. *J.R.S.S. (B)*, **20**, 334–343.
Wilks, S. S. (1963) *Mathematical statistics*. Wiley, London.
Williams, W. T., Lance, G. N., Webb, L. J., Tracey, J. G., and Connell, J. H. (1969) Studies in the numerical analysis of complex rain-forest communities. IV: A method for the elucidation of small-scale forest pattern. *Journal of Ecology*, **57**, 635–654.
Wishart, D. (1969a) Numerical classification methods for deriving natural classes. *Nature*, **221**, 97–98.
Wishart, D. (1969b) Mode analysis. In *Numerical taxonomy*, Ed. A. J. Cole. Academic Press, New York, pp. 282–308.
Wishart, D. (1969c) An algorithm for hierarchical classifications. *Biometrics*, **25**, 165–170.
Wishart, D. (1978) *CLUSTAN user manual*. Program Library Unit, Edinburgh University.
Wolfe, J. H. (1969) Pattern clustering by multivariate mixture analysis. *Multivariate Behavioural Research*. **July 1969**, 329–349.
Young, T. Y. and Calvert, T. (1974) *Classification, estimation and pattern recognition*. New York, American Elsevier.
Young, T. Y. and Coraluppi, G. (1970) Stochastic estimation of normal density functions using an information criterion. *IEEE Transactions on Information Theory*, **IT-16**, 258–263.
Young, T. Y. and Farjo, A. A. (1972) On decision-directed estimation and stochastic approximation. *IEEE Transactions on Information Theory*, **IT-18**, 671–673.

Index

Applied Probability and Statistics (Continued)

FLEISS • Statistical Methods for Rates and Proportions, *Second Edition*
FRANKEN • Queues and Point Processes
GALAMBOS • The Asymptotic Theory of Extreme Order Statistics
GIBBONS, OLKIN, and SOBEL • Selecting and Ordering Populations: A New Statistical Methodology
GNANADESIKAN • Methods for Statistical Data Analysis of Multivariate Observations
GOLDBERGER • Econometric Theory
GOLDSTEIN and DILLON • Discrete Discriminant Analysis
GROSS and CLARK • Survival Distributions: Reliability Applications in the Biomedical Sciences
GROSS and HARRIS • Fundamentals of Queueing Theory
GUPTA and PANCHAPAKESAN • Multiple Decision Procedures: Theory and Methodology of Selecting and Ranking Populations
GUTTMAN, WILKS, and HUNTER • Introductory Engineering Statistics, *Second Edition*
HAHN and SHAPIRO • Statistical Models in Engineering
HALD • Statistical Tables and Formulas
HALD • Statistical Theory with Engineering Applications
HAND • Discrimination and Classification
HARTIGAN • Clustering Algorithms
HILDEBRAND, LAING, and ROSENTHAL • Prediction Analysis of Cross Classifications
HOEL • Elementary Statistics, *Fourth Edition*
HOLLANDER and WOLFE • Nonparametric Statistical Methods
JAGERS • Branching Processes with Biological Applications
JESSEN • Statistical Survey Techniques
JOHNSON and KOTZ • Distributions in Statistics
Discrete Distributions
Continuous Univariate Distributions—1
Continuous Univariate Distributions—2
Continuous Multivariate Distributions
JOHNSON and KOTZ • Urn Models and Their Application: An Approach to Modern Discrete Probability Theory
JOHNSON and LEONE • Statistics and Experimental Design in Engineering and the Physical Sciences, Volumes I and II, *Second Edition*
JUDGE, GRIFFITHS, HILL and LEE • The Theory and Practice of Econometrics
KALBFLEISCH and PRENTICE • The Statistical Analysis of Failure Time Data
KEENEY and RAIFFA • Decisions with Multiple Objectives
LAWLESS • Statistical Models and Methods for Lifetime Data
LEAMER • Specification Searches: Ad Hoc Inference with Nonexperimental Data
McNEIL • Interactive Data Analysis
MANN, SCHAFER and SINGPURWALLA • Methods for Statistical Analysis of Reliability and Life Data
MEYER • Data Analysis for Scientists and Engineers
MILLER • Survival Analysis
MILLER, EFRON, BROWN, and MOSES • Biostatistics Casebook
MONTGOMERY and PECK • Introduction to Linear Regression
NELSON • Applied Life Data Analysis
OTNES and ENOCHSON • Applied Time Series Analysis: Volume I, Basic Techniques
OTNES and ENOCHSON • Digital Time Series Analysis
POLLOCK • The Algebra of Econometrics
PRENTER • Splines and Variational Methods
RAO and MITRA • Generalized Inverse of Matrices and Its Applications
RIPLEY • Spatial Statistics
SCHUSS • Theory and Applications of Stochastic Differential Equations
SEAL • Survival Probabilities: The Goal of Risk Theory
SEARLE • Linear Models
SPRINGER • The Algebra of Random Variables
UPTON • The Analysis of Cross-Tabulated Data
WEISBERG • Applied Linear Regression
WHITTLE • Optimization Under Constraints